OLIVE OIL AND VINEGAR FOR LIFE

OLIVE OIL AND VINEGAR FOR LIFE

Delicious Recipes for Healthy Caliterranean Living

THEO STEPHAN

Skyhorse Publishing

www.skyhorsepublishing.com

Library of Congress Cataloging-in-Publication
Data available on file.
Print ISBN: 978-1-5107-0653-8
Ebook ISBN: 978-1-62873-182-8

Cove design by Jane Sheppard

Printed in China

GLOBAL GARDENS®
This book features multiple
award-winning extra virgin olive
oils & vinegars from Global
Gardens, Los Olivos, California,
Santa Barbara County

For HELEN;

YOU KNOW WHY!

TABLE OF CONTENTS

THE INGREDIENTS

THE LIFESTYLE

CELEBRITY CHEF FEATURES

THE RECIPES

OLIVE OIL

IN THE BEGINNING . . .

In ancient Greek mythology the goddess Athena planted an olive tree as a gift on the rocky hill now known as the Acropolis—and tour guides claim that the tree standing there today comes from the very same roots. Most historians agree that the island of Crete was the location of the first cultivation of the fruit, with the earliest olive oil amphorae (ceramic containers) dating to around 3500 BC, although archaeologists have dated ancient olive oil presses to 5000 BC. The first actual recording of olive oil extraction is found in the Hebrew Bible—the hand-squeezing of the fruit dating to the thirteenth century BC during the Exodus from Egypt. Gladiators rubbed their bodies with it, and the winner's body was then scraped—the resulting sweat and olive oil packed into tiny amulets and then sold, supposedly bringing the bearer strength and longevity. Homer was the first writer to refer to the cherished oil of many uses as "liquid gold."

California olive oil production is relatively new when compared to these origins. Spanish missionaries began carrying olive trees (along with grapevines for wine) to the New World in the 1500s, but commercial olive oil production in the state wasn't recorded until around 1870. Growing interest in olive oil's health benefits and the ability to grow olives in certain regions of the state encouraged contemporary production in northern California during the '70s. Similar to wine production, the trend drifted to Southern California some twenty years later.

My first experience with olive oil came at the ripe age of eight, watching my favorite aunt Lou fry eggplant to layer in her famous moussaka. I asked her why her eggplant tasted so good and my mom's (sorry, Mom!) didn't. She reached up and pulled a Wonder Bread hamburger bun out of a bag—it was 1968, the height of processed food popularity! She poured some olive oil from a large tin with Greek letters on it into a little cup, handed me the bun, and said, "Here, dip the bread in *this.*" My mouth immediately welcomed the buttery, fresh, flowery fruitiness that is Kalamata Extra Virgin Olive Oil. She then poured some Crisco oil into another bowl and said, "Now taste *this*!" I made a terrible face. She laughed and said, "*That's* what your mom uses!"

The genesis of my own journey into the art of growing, harvesting, pressing, and bottling olive oil began in total ignorance. I was walking down a side street in Santa Barbara, California, in the mid-'90s, feeling smug about a graphic design contract I had recently acquired in Hollywood. Above me was a brilliant blue California sky. Beside me was a majestic spindly olive tree. Strong, well-groomed branches reached into the warm, dry air, opening their fruit-laden arms to the sky. It was autumn and I noticed some ripe black olives had fallen to the ground. Like a kid stealing candy, I took a quick gander around—no one was looking. I snatched up the most perfect olive, wiped it off on my jeans, and popped it into my mouth. Oh, the impossible viscosity, bitterness, and downright poisonous flavor that permeated my palate! Needless to say, I made a spectacle out of myself spitting it out onto the sidewalk. I am happy I didn't know anybody in Santa Barbara at the time . . . it's a fairly small town!

Then I got inquisitive—what made olive oil taste so good if the fruit that created it tasted so bad? Until that very moment, I hadn't given it a second thought. I bet my aunt Lou would have known!

At that time I had just invested in a small ranch where I could work from home, commuting to Los Angeles and back to Ohio when needed (my graphic design firm

Real Art Design Group was based in Dayton, Ohio). A fifty-acre fixer-upper ranch in Los Alamos, California—in the northwestern part of Santa Barbara wine country—had needed a buyer and I had needed a home. This became the original planting for Global Gardens (the first in Santa Barbara County and all of Southern California specifically for certified organic extra virgin olive oil).

While I loved graphic design, after my first entrepreneurial endeavor (selling American Seeds door-to-door as a six-year-old) I longed to create my own edible garden, taking advantage of the microclimates and vast temperature changes of the Santa Ynez Valley, while perfecting the chemistry of dirt and water. On a snowy winter day back in Dayton, Ohio, the eureka moment came—why not start my own brand, grow and sell olive oil from that handsome property, and quite possibly, daringly change my life forever? No more Four Seasons Beverly Hills. No more maniacal driving to the Burbank Airport to receive 6:00 a.m. flights incoming with that day's presentation on board. No more 1:00 a.m. press proofs. Sounded great to me. My love for graphic design, fine arts, gardening, food, entertaining, and, yes, even science could be combined into one new business. I had been creating and selling brand concepts for years . . . why not create one for myself?

I quickly took an olive propagation course at UC Davis, where I learned a plethora of olive knowledge that my ancestors would have been proud of—including growing techniques, budgeting, and how to professionally assess olive oil flavor nuances. Everything seemed to have happened at about the same time, including the opportunity to adopt two beautiful girls (biological sisters) from Nepal. This made the career switch even more meaningful—to create an environment of nurturing and an agricultural yield of meaningful abundance. Anita and Sunita took part in our first harvest of 2001, and continue to celebrate the tradition now. That same year, my acquired partners from Real Art sent me my last payment, purchasing the firm in full. I was free—or was I?

Traveling the world was always a great passion of mine—tasting foods and flavors on every continent except Australia and Antarctica. (I'll get there!) Exotic species of spices, fish, fowl, vegetables, fruits, nuts, and sometimes even things more untraditional were (and still are!) welcome on my palate. My new brand, Global Gardens, became the way to memorialize those flavors, created with both intention and attention in California.

The year 2016 marks my sixteenth harvest year and my eighteenth year formulating fruit vinegars with no added sugar; the sugar content comes from the fruit itself! One of the unique aspects of Global Gardens bottle labeling is that I insist we be varietal specific. You won't find this distinction very frequently on the olive oils you buy at your grocery or specialty food store. Why? Good

Theo giving a tasting at her Global Gardens store

question. Just like wine, the nuances of the olive fruits themselves, times and styles of harvests, along with stainless steel hammer mill versus stone milling options, make a huge difference in taste. Global Gardens' customers have come to appreciate the difference and rely on our labeling to further their olive oil education!

My original olive trees and the regional groves I manage are hand-harvested, crated, and pressed by varietal. Thirty pickers gather at dawn, around 6:30 a.m., on a late autumn day of my choosing—when the olives are just the right color. For some varietals, that means green, others purple, the rest black. You've never seen a black martini olive, right? That's because the sevillano olive, which makes the best martini olive, is at its prime when it's plump and green. Left on the tree to turn black, it will get mushy. Green kalamatas? No way!

That is an olive varietal that is ripe when it is anywhere from a light purple to dark purple. People refer to olives as "black," but they're actually the deepest hue of purple. When you see a light or medium brown canned olive, it was actually purple when harvested. The brown color comes from the curing process itself.

Read on for insight about how Global Gardens oils are made from the very start from olives to extra virgin, first cold pressing perfection.

Within twenty-four hours of harvesting, the olives must be pressed to maintain low acidity in the fruit and the highest flavor points. (Ever eaten corn on the cob just minutes after harvesting compared to days after? Incomparable! When produce is freshly harvested, the natural sugars are pumping through the entire plant, releasing into the ripest fruits. If it's a long time until they are eaten, or if they are harvested prior to their prime, they really don't taste good at all).

Harvest begins at 6:00 a.m., typically still before the fog burns off. The olives await their ultimate purpose, (1) picked by hand (2) and crated in air-flow-thru one-half-ton bins.
(3) The picking process is laborious and lasts until nightfall. The closest certified organic stone mill is almost six hours away. The olives are tucked in for a good night's sleep and driven to the mill after high-traffic hours, arriving just before dawn (4), unloaded, and weighed (5). Next they are emptied bin by bin into an open metal container (6) that has a metal screening, small enough to catch twigs and

1) The olives are picked by hand.

2) The olives are crated.

3) Laborious for some, that is.

leaves undesirable in the milling process, large enough for multiple olives to gently drop through and travel up a ladder-style conveyer (7) that shimmies out any remaining vegetation, then through a gentle tumble washer (8).

The key word here is "gentle." Olives that are bruised—as with any bruised fruit—will have a higher acidity level from that bruise (a.k.a. rot). Olives need to be treated especially well during and after harvest. We

never pick up loose fruit that has voluntarily found its way to the ground. The one-half-ton bins may appear small; however, the weight of the fruit itself would crush and bruise the fruits at the bottom if those bins were any larger.

A conveyer belt (9) delivers the olives to a twelve-foot stainless steel bowl containing two large stones (10) that not only revolve, but also are connected by an axle that spins the set of stones clockwise as they are

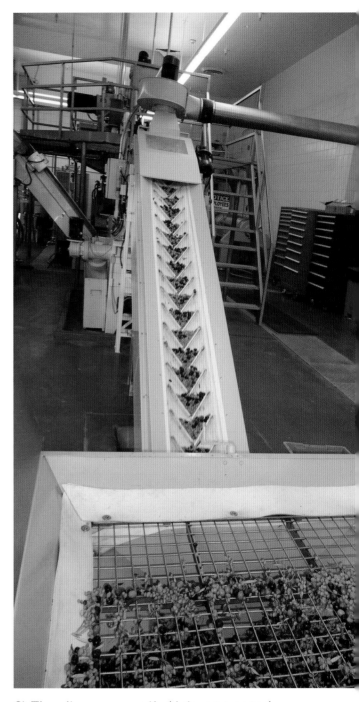

4) The olives arrive before dawn.

5) The olives are unloaded and weighed.

6) The olives are emptied into a screened metal container that catches twigs and leaves.

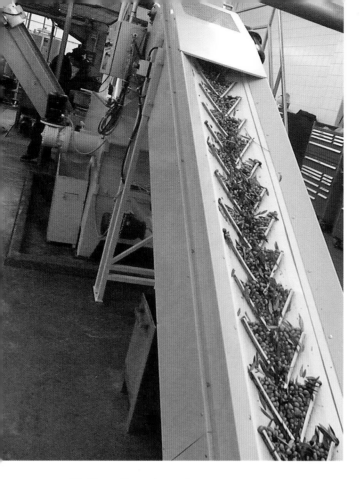

7) The olives travel up a conveyor . . .

8) and are gently washed.

9) Along another conveyor belt the olives are delivered to a steel bowl.

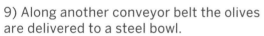

10) Inside the bowl are two large revolving stones that mash the olives.

12) A centrifuge spins the mash to separate the oil and water.

11) A stainless steel mixer mashes the remaining fruit and seed fragments.

turning and mashing the olives. Most of the oil (as in most essential oils also) derives from the seed of the olive. Surprised? Me too. I wonder *who* exactly, in ancient Greece and Rome, had the prescience to want to burst open that seed, knowing the most precious of liquid gold awaited them for their labor.

A stainless steel horizontal mixer (11) blends the mashed fruit and seed fragments thoroughly. A contained centrifuge (12) spins the mash while simultaneously pressing on it. This allows the water to separate from the oil. The resulting green luscious love permeates the entire industrial pressing room with an aroma too beautiful to describe. We run for fresh bread and wine, enjoying the relief and flavors of our labor.

This book features many Global Gardens varietal-specific extra virgin olive oils and vinegars. Sure, I'm partial to my own brand, but I enjoy trying every olive oil shared with me, and I particularly love to visit other olive groves around the world, appreciating sometimes subtle, other times dramatic differences that same varietal olive oils create on our palates. For example, I grow Greek koroneiki trees, imported from the island of Crete. My Koroneiki Extra Virgin Olive Oil tastes completely different from the koroneiki my friend Yiorgos Demitriadis produces from his Biolea groves, which produce their own award-winning oil on the primeval island.

A few words about baking with extra virgin olive oils: I prefer the buttery varietals like kalamata, koroneiki, and Mission. *However*, the koroneiki extra virgin I import from Crete tastes totally different from the koro I grow here in the States, and the Global Gardens Mission, grown up in Central California, is much more buttery than the Mission from our own groves too. This all said, there is really no substitute for my sagey-finish Mission/ Manzanilla blend called for in my Perfect

Pecan Pie (page 215) that was featured in the *Los Angeles Times*. The savory finish seems in perfect symbiosis with the nuts and maple syrup in the recipe. Just be a bit forewarned, adventure prevails! When you substitute olive oil for butter, remember: Premeasuring in one-tablespoon increments and then freezing olive oil in ice cube trays helps it to act a bit more like butter, and always start with less olive oil than the butter in your traditional recipes. You can always add more, but you can't take it away!

I have spent almost twenty years developing varietal extra virgin olive oils, fruit vinegars, and people-pleasing products. My olive oils have won over twenty medals and are one of the most awarded in Southern California. The LA County Fair is the world's largest and most prestigious competition; we've won every year we've entered! The recipes in this book comprise over thirty years of trial and error— pairing them with different varietal olive oils to see which works best. The particular olive oils and vinegars spelled out here are my own favorite flavor combinations. Every palate is different, so I encourage you to find your own taste bud tantalizers. For me, cooking isn't about pretentious terms, difficult tasks, or unshared secrets. Ultimately, it's about *eating* and enjoying the multitude of available fresh flavors we have in our local and global marketplace, combined with the freshest, healthiest extra virgin olive oil . . . for life!

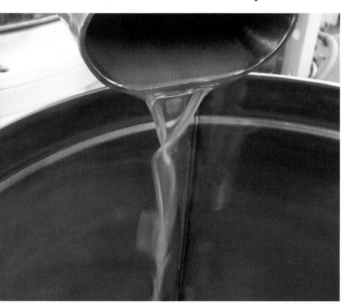

WHAT'S AN EXTRA VIRGIN?

Extra virgin olive oil *should be* defined as containing no more than 0.8 percent acidity—it has a superior taste, color, and smell. Olive oil flavor descriptions can range from a generous list of fruit varietals to grassy and peppery, along with an immense range of herbs. The significance is that in a real extra virgin olive oil, one, if not several, of these flavor profiles will be apparent.

I say "should be" defined because, sadly, many of the olive oils found on grocery store shelves do not *test* as extra virgin. A study published by reputable research authors at the UC Davis Olive Center details their analysis of many popular brands of "extra virgin" olive oil and found 69 percent of imported and 10 percent of California extra virgin olive oils were, in actuality, not. For more information about the massive fraud occurring in the olive oil industry, check out *The New Yorker*'s fascinating feature in the August 13, 2007, issue.

The truth is found in the flavor. Many folks come into the Global Gardens Destination Store in downtown Los Olivos, California, with the perception that olive oil should not have any taste at all. Educating customers on the contrary is our greatest joy. Varietal bottling from Global Gardens captures the distinctive *terroir* of its grove of origin,

very similar to wine grapes. A pinot noir from Washington state tastes completely different from a pinot from Santa Barbara County. Diverse fruitiness and peppery finishes ranging from mild to robust—again, dependent on the varietal and geographical location and microclimate (we have four different temperature zones on one fifty-acre olive ranch)—will be represented in a real extra virgin olive oil.

Maintaining the quality of olive oil is another factor. Always store your extra virgins in a cool, dry pantry—never in sunlight, above the kitchen stove, or in the refrigerator. I go through a bottle of olive oil frequently, so I put capped bottle pours on their tops and place them on my counter for convenience. If you're a frequent user like me, the oil won't go bad in the two weeks it takes to devour it! Refrigerating olive oil only makes it coagulate, ruins the flavor, and requires bringing it back to room temperature for use. One customer proudly told me she brought her coagulated olive oil back to life by heating it in the microwave. Uh, yikes!

I always travel with at least one bottle of Global Gardens Extra Virgin Olive Oil. Call me weird, but I love to cook when I'm on vacation and try to rent houses instead of staying in hotel rooms. On a recent excursion I was aghast to discover everything *but* olive oil in my traveling kitchen bags. I went to the nearest store and bought the most expensive California extra virgin olive oil they had—a frequent award winner and one I knew. I was excited to try it. As soon as I opened it, I was met with an unfortunate peanuty aroma. I took it right back to the store. I'm convinced that the rancidity of the oil was caused by photooxidation. This can happen under halogen and store lights, causing serious deterioration to the natural chemical balance of the oil. I encourage you to be very particular about your oils. When you buy one that is not right—no matter the cost—you should ask for your money back at your local food purveyor if it is indeed labeled *extra virgin olive oil*.

First cold pressing means that no heat has been added to the extraction process that, again, affects flavor dramatically. Adding heat also affects yield—the warmer oil is thinner and will generously increase harvest results. But not at Global Gardens! All of our oils are extra virgin first cold pressing, thank you very much.

Typically, the darker the fruit harvest, the less bitter the oil, especially when pressed on a stone olive mill versus a stainless steel hammer mill press. Stainless steel tends to sharply slice into the fruit and seed, running hotter (because it's metal) than a stone mill, with the heat contributing to bitterness. Most of our oils are run on a stone mill, with select oils coming from stainless steel mills we trust not to run

too hot. Bitterness can also come from the varietal and harvest as well as the pressing style—but a great extra virgin olive oil should not be extremely bitter, no matter where it is from. A few of our Global Gardens California certified organic extra virgin olive oils have a mildly bitter finish—and culinary bitterness can be a good thing in moderation. These oils are perfect in those dishes where you might add rosemary, basil, or other strong herbs—as opposed to using as dipping oil or for mild dishes. Most of the oils featured in this book have a pleasant aftertaste and are described on pages 48 through 50.

I've encountered a few wary cooks that were hesitant to try cooking with extra virgin olive oil. Why? I want my family to have the tastiest and most healthy essential fats in their diet (see next section on nutritional benefits), and extra virgin olive oil is the perfect oil for all cooking and baking uses. I haven't found a recipe yet that did not benefit from the flavor, texture, and aroma of olive oil! The smoke point of extra virgin olive oil is 385. You can even deep-fry with it! When olive oil does smoke, it's probably because it's been adulterated with a lesser oil of a lower smoke point. Other reasons might include that your pan is too hot or perhaps you're using the wrong type of pan. Deep-frying in a wok, for example, makes it fairly impossible to control the temperature. I use an electric deep fryer that is temperature controlled.

NUTRITIONAL BENEFITS YOU NEED TO KNOW

Ever wonder how you can exercise and starve yourself on vegetables and tuna fish packed in springwater but never lose weight—or if you do, you gain it right back? The solution lies in the fact that our bodies require healthy and essential fats for us to lose weight and *keep it off*. Olive oil is a monounsaturated fat and can be a phenomenal health attribute—don't look at the calories and shove the bottle back in the cupboard! The monounsaturated fatty acids in olive oil help to lower the risk of heart disease in many people by improving risk-related issues such as cholesterol levels. Some research even shows that type 2 diabetes patients can enjoy better blood sugar control and better insulin levels by using moderate amounts of olive oil in their diet.

When we use extra virgin olive oil on pasta, bread, potatoes, or other carbohydrates, the chemistry in our body slows down the absorption rate of the carb during the digestive process. This allows the carbohydrate to maintain its healthy properties instead of quickly turning into sugar, which happens to most carbs. This is why pastas and potatoes get a bad dietary rap when, in actuality, the problems begin with what we *put* on the pasta and potatoes! This doesn't mean you can pour olive oil on a pound of pasta and fill up to your eyeballs. It *does* mean that you can enjoy most carbohydrates in moderation (see serving suggestion on packaging for proper amount) with a measured amount of extra virgin olive oil. I made sure that the recipes in this book offer healthy quantities per portion.

Extra virgin olive oil contains a high amount of polyphenols that are natural antioxidants. Countless health studies point to a reduction in risk of coronary disease, blood pressure, and lower cholesterol counts when our bodies absorb healthy amounts of phenolic compounds. Studies indicate these compounds can reduce the inflammation that causes coronary heart disease. Decreasing the most damaging form of cholesterol, LDL,

CREDIT: ROBERT DICKEY

is another powerful effect found within the antioxidant power of extra virgin olive oil. Ongoing studies note that compounds from the olive fruit were found to be antimicrobial against various undesirable bacteria. The consumption of extra virgin olive oil has been known to improve the sensitivity of cells to insulin, which helps to decrease the risk in many patients of cardiovascular disease, diabetes, obesity, and Alzheimer's disease.

The koroneiki olive varietal has been found to contain the highest level of polyphenols. Global Gardens sells an imported koroneiki from the island of Crete that is super fruity and buttery, perfect for taking off mascara and using as a natural skin moisturizer, I do! Men can get great skin benefits by replacing their shaving cream with Koroneiki Extra Virgin Olive Oil. Global Gardens also harvests koroneiki from our own California certified organic groves, producing a limited amount each year, highly prized for its strong, healthy attributes. The terroir surrounding this grove includes acres of California native sage, so our Koroneiki tends to be quite robust and perfect for cooking with a sagey finish.

Surprising to many, one tablespoon of extra virgin olive oil contains 8 percent of our recommended dietary allowances (RDA) of vitamin E. Olive oil is also the second best source of vitamin K, the richest sources found in dense, leafy vegetables like kale and broccoli. Vitamin K is necessary for our metabolism to process proteins, creating health for density in bones and tissue.

FLAVOR PROFILES AND HOW TO PAIR THEM WITH FOOD

Olive oil pairing when cooking? You bet! You pair different wines with foods, so why not olive oils, they have a lot in common with wine. *Terroir*, derived from the French word that means "land," describes the particular characteristics imparted onto the flavor of a fruit—most commonly coffee, wine grapes, and tea. Climate, geography, topography, and geology all play certain roles in developing these essential terroir personalities. Terroir matters when it comes to the ultimate flavor of an olive oil.

On page 16 is a simple chart of some of the most common flavor profiles associated with extra virgin olive oils. The descriptions on the left are flavors you *want* to taste.

The list on the right represents flavors that would indicate an old, rancid, or adulterated olive oil:

The first time I went to a formal wine tasting I was really pleased when I could actually make out the taste of the chocolates and blackberries in the latest Syrah. Now that you have some flavor notes, have fun! Pull the olive oil(s) out of your cupboard and put one-fourth teaspoon onto the middle portion of your tongue. Swallow. See what kind of sensations you get. If you get a tingly, peppery feeling in the back of your throat, in addition to any of the flavors on the good flavors list, it's likely that your olive oil is certainly an extra virgin. If not, well, I'd dump it out and find one that is—*you deserve it!*

Check out our olive oil tasting kit available at the Global Gardens online store or at our Los Olivos, California, destination store. You will receive a Global Gardens signature tasting tray and specific instructions on how to have your own tasting party— whether it's by yourself or with a group of friends and family, a tasty, educational experience offered exclusively by Global Gardens awaits you.

You can readily see by the descriptions on the Good Flavors list that you might want to use the fruity-flavored oils in baking or desserts, the herbaceous on pastas, peppercorns on salads, the woodsy, more robust flavor profiles in stronger dishes like salmon or leg of lamb, etc. Chocolate brownies and cake mixes? Throw out your canola oil, people! Global Gardens Kalamata, Koroneiki, Manzanilla, Mission (the buttery one), and Ascolano Extra Virgin Olive Oils each make for the perfect "vegetable oil" called for in commercially prepared mixes. Now that you have a simple guideline, it will be interesting to do your own experimenting. For this reason, I label all Global Gardens extra virgin first cold pressing oils with the varietal name. If they are blends of two or more varietals, I name them all. Knowing the varietal name will help guide you in food pairing as you discover your own particular favorites. But remember, my 100% Mission Extra Virgin Olive Oil may taste totally different from the 100% Mission Extra Virgin grown in a different part of the state. Enjoy the differences and arm yourself with a little education. Our website has flavor descriptions of each varietal we grow and sell. You can get a great start by reading more there.

GOOD FLAVORS	BAD FLAVORS
Apples	Bland
Fresh almonds	Blue cheese
Artichokes	Brine
Bananas	Burnt
Butter	Flat
Butterscotch	Fusty
Cinnamon	Frozen
Fruity	Greasy
Grassy	Heavy
Herbaceous (any fresh herbs, especially rosemary)	Metallic
Mangoes	Resin
Melon	Vinegary
Pears	Smoky
Peppercorns (green, white, pink, black)	Stale nuts
Spicy	Tonic
Tomato plant	Mushrooms
Tropical	Musty
Woodsy	Yeasty

YOU DON'T ALWAYS NEED THE VINEGAR

The first time I went to Greece, I ordered my favorite salad—the traditional horiatiki—fresh cucumbers, garden red tomatoes, onion, kalamata olives, feta cheese, a sprinkling of Greek oregano, sea salt, and a heavy drizzling of olive oil. When the proprietress came to clear the table—meeting the owner at the end of the meal is common in small island tavernas—she asked if something was wrong with my salad. I had eaten almost every morsel. I was famished from a sunny day of swimming and zigging around curvy roads in an open-air Suzuki. Perplexed by her question, she pointed to the pool of olive oil left on my plate and said, in Greek, eyes twinkling, "Didn't your grandmother teach you better?"

Now that I know what it takes to make a gallon of this precious, sumptuous liquid gold, I better understand the question! It takes between forty and fifty pounds of olives to make just one gallon of extra virgin first cold pressing olive oil. The owner of the taverna was right—and now I say the same thing to my guests! Olive oil—when it's really *real*, fresh, extra virgin olive oil is a commodity to be savored and appreciated—it's a perfect stand-alone salad dressing, pasta sauce, or grilling glaze. Add fresh herbs and minced garlic and try it. Unpretentious, fresh, and, perhaps best of all, easy. The best part is that your metabolism, heart, stomach, and skin will actually thank *you*.

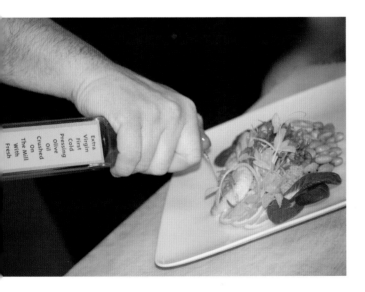

VINEGAR

ORIGINS AND INTERESTING FACTS

Ranch is out. Thousand Island is for Big Macs . . . Italian and French? Those are for people who aren't worried about the dollar versus euro exchange rate. Vinegars in all flavors are making a quick entry into chefs' kitchens and everyday pantries as we move beyond traditional balsamics and opt for delicious fruit vinegars that conjure memories of strawberry fields forever and tropical vacations.

CREDIT: KAREN FAHDEN

The word *vinegar* actually comes from the French word *vinaigre*, meaning "sour wine." Its uses and offerings go back thousands of years: Hippocrates is said to have prescribed it for certain ailments, Helen of Troy bathed in vinegar to relax, Caesar's armies drank vinegar to avoid dehydration, and medieval citizens rubbed vinegar over their skin as a germ protectant during the bubonic plague. The Babylonians are the first to have actually used vinegar as a preservative for pickling food.

Ah, the science of food! Vinegar begins its tartish life as another French phrase, *mere de vinaigre*, which literally translates into "mother of vinegar." More simply referred to as "mother" in English, it is a gooey substance composed of various bacteria that cause the fermentation in wine, fruits, ciders, rice, etc., and evolves them into vinegar. Vegetables, grains, fruits, and juices from almost anything starchy or sweet can be transformed into vinegar. The process

itself isn't very romantic (or photogenic!). Basically, the mother, along with the fruit juice, is brewed in megagallon stainless steel tanks until all the alcohol is cooked away. The vinegar is then aged in oak barrels for about a year.

THE QUEST FOR
BALSAMIC CONQUEST

The original Italian balsamic vinegar is made from white Trebbiano and/or Lambrusco grapes grown only in Modena, Italy. Grapes are crushed the old-fashioned way—on wooden hand-cranked presses—filtered, boiled, cooled, filtered again, then aged in vintage barrels for a minimum of twelve years. The longer it ages, the sweeter and more syrupy it becomes, but that also means the sugar/calorie content is much higher.

Global Gardens' Balsamic Fruit Vinegars are quick-brewed by a professional vinegar maker in California. I met Karen Fahden shortly after planting my olive grove, in 1998. I had just completed a season of harvesting abundant amounts of raspberries, big as my thumb, sweeter than sugar, making about forty cents an hour selling them to local markets—discouraging work! Karen suggested infusing my

succulent sweethearts into her vinegars, and the rest is history. I have developed a formula that vastly outperforms commercial fruit vinegars. Our golden balsamics originate as chardonnay wines, and the dark balsamics are made using merlot/cabernet blends for rich, consistent, and heavenly fruit infusions! The sugar content on our labels comes from the natural sugars in the fruit itself, usually only 4 grams per tablespoon.

In our technologically advanced twenty-first century, chefs and culinary buffs are constantly discovering and inventing cool fusion flavors cooked up in a lab instead of nature, but they can't hold a candle to the natural delicious taste of vinegar in your cooking. Give yourself a break from your tedious reliance on the traditional, plain balsamic, or (horrors!) processed commercial salad dressings—and get adventurous with the many other options out there. Or give the original stuff a try. The real Italian balsamic vinegar should contain the words "aceto balsamico

tradizionale" on the bottle, and it would have been aged for at least ten years— some even for fifty or even one hundred years. Authentic balsamics are expensive due to their tedious production. Truly to be savored.

BEYOND NUTRITION

Does vinegar really burn fat? According to several studies, it does! This concept finds its origins in the *Journal of Nutrition*, in the first vinegar study of scientific significance by Nobumasa Ogawa in the year 2000, which dealt with blood glucose levels. Arizona State University professor Dr. Carol S. Johnston discovered, quite by accident in 2004, that her subjects who took two tablespoons of vinegar before lunch and dinner lost an average of two pounds during a four-week period during the holiday season, a time when people typically eat more than usual. (She was researching cholesterol levels, not weight loss.) Of even more interest is a 2009 study completed by Tomoo Kondo, resulting in mice using a high-fat diet (that included vinegar) developing less body fat than mice who did not ingest vinegar. The explanation that the research scientists agreed on was the fact that the acetic acid in the vinegar promotes body-producing proteins that break down fat and discourage fat buildup in the body.

Research has shown that adding vinegar to a meal slows down the rate at which carbohydrates metabolize into your bloodstream (a.k.a. sugar conversion). Vinegar improves the absorption and utilization of essential nutrients, like calcium, that are locked in foods, benefitting our overall digestive system. Dietary calcium, especially important in preventing osteoporosis, is difficult to digest in supplements, especially for lactose-intolerant women. Kale, chard, and other dark leafy greens are vitamin packed, but they contain compounds that actually inhibit calcium absorption. A few drizzles of tangy vinegar may change all that!

Super tip for sodium watchers: if you're cooking beans—in any dish—add some vinegar near the end of the cooking process. It will dramatically decrease the amount of salt that beans require, raising the flavor quotient without raising your blood pressure!

There are a lot of "miracle cures" in the ever-expanding world of hyper information. But rest assured that many studies show vinegar can help improve your general health in many ways.

CREATIVE CULINARY USES AND STORAGE

Suppose you replace lemon, called for in any recipe, with Global Gardens' Apricot, Pomegranate, Mango, Meyer Lemon Balsamic Bliss, Blood Orange, Strawberry, Fig, Raspberry, Passion Fruit, or Green Apple with Ginger, Balsamic Fruit Vinegars. (I'm always working on new flavors, so check our website for the latest and greatest!) Try replacing one-third of the water with your favorite Global Gardens fruit vinegar when making rice—my favorites for rice are the Pear with Tamarind, Mango, Passion Fruit, and Fig Balsamic. Always remember to shake the all-natural, fruity goodness from the bottom of the bottle before using.

Adorn your next meal with a reduction made by simply bringing the fruit vinegar to a fast boil, then watching it thicken, reducing to about 50 percent. Drizzle this simple but exquisite concoction over veggies, Brie cheese, seafood, chicken, pork, beef, gelato (anything!) for a new sweet-and-sour tangy thrill. There is such a mystique about reduction glazes when, honestly, they take less than fifteen minutes and a bit of vigilant stirring so they don't stick to the bottom of your pan. See page 27 for a simple, quick Global Gardens Fruit-Infused Balsamic Reduction recipe.

Add my fruit-infused vinegars (a quarter cup per two quarts) to water when you are parboiling ribs or meat for stews; even the toughest meats will become fork-cuttingly tender! Homemade breads and rolls will shine with playful twinkles when brushed just before they are finished baking. Make a cake super moist by adding a tablespoon of, say, pomegranate vinegar to chocolate cake mix—or mango, strawberry, raspberry, fig.

Now, take a can of tuna fish. Boring? No more! Use two tablespoons of your favorite fruit-flavored vinegar, add capers, cilantro, boiled egg, salt, and fresh ground pepper . . . even some chunks of the fruit itself if you have it. Voila—nicoise salad transforms before your pupils by adding boiled redskin potatoes and some farmers' market salad greens.

Vinegar can be used in place of mayonnaise, ketchup, and lots of other calorie-laden, cholesterol-producing flavors we really don't need in our cupboards. Fish and chips (for fried foods, see Caliterranean Frying, pages 156–159), coleslaw, salad dressings, and sandwiches will all gain flavor and beneficial health aspects when you use one of my vinegars, enhancing your palate and your life at the same time.

Fruit vinegars should be stored in a cool, dry place. Mine are packaged in pretty bottles, so I keep them on the counter and they get used up in no time, staying super fresh if not exposed to direct light. Most commercially produced fruit-infused vinegars are filtered—or put into dark bottles. I use clear glass and nonfiltered fruit in my fruit-infused vinegars because the colors are simply sumptuous, reminding me of their fresh beginnings. Shaking them up once

in a while (not necessary if you use them once a week) will keep them fresh; I do not suggest putting them in the refrigerator since the acid from the vinegar will maintain the stabilization—and I loathe the thought of them being forgotten amongst the mayo, ketchup, and your great aunt's pumpernickel chutney. If you're not going to use them or shake them with a little dance move, you will have to refrigerate them or else the mother may form. You will know if that happens when you see a balloonish blob swimming in your vinegar—that means it's time to start fresh again! Right now, the beloved cerignola and kalamata olives from my own backyard would actually keep indefinitely (if we didn't gobble them up so fast!), each jar covered with a different Global Gardens Balsamic Fruit Vinegar and refrigerated.

Just imagine a Blood Orange Dark Balsamic Vinegar oozing over green olives sprinkled with fresh herbs and honeyed orange rinds. Using fruit vinegars as culinary enhancements has endless possibilities; I hope to hear from you with your own creative input.

THEO'S OLIVE CURING RECIPE

Choose firm, dark olives; if you have green olives mixed into the batch, the brining may take a week or two longer. Put in a clean glass or ceramic jug. If you have a lot of olives, drill ¼-inch holes in the bottom of a plastic, food-grade 5-gallon bucket, all around, like a strainer. Use another bucket (don't drill holes in the second bucket) to hold the strainer bucket filled with olives.

Make a solution with a ratio of ¼ cup salt dissolved into 1 quart water and pour enough over the olives to cover, then weight the olives with a stone or inverted plate so they are completely submerged. If you're working with large buckets, partially fill a tall kitchen trash bag with water to keep the olives submerged.

Store in a cool place, changing the solution *once a week for 6–12 weeks, depending on varietal and ripeness*. If a scum forms on the surface during that time, disregard it until it is time to change the brine, then rinse the olives with fresh water before covering with brine again. The scum is harmless. At the end of four weeks, taste one of the largest olives. If it is only slightly bitter (these olives should be left with a bit of a tang), pour off the brine and rinse the olives. If the olives are too bitter to be put in the marinade, rebrine and soak for another week (or more, to taste), then rinse and marinate.

BE CREATIVE, JUST USE THESE PROPORTIONS!

One cup Global Gardens Balsamic Fruit Vinegar (Martha Stewart herself loved my Pomegranate Balsamic Kalamatas at a party I catered in Palm Springs. I personally saw her enjoy several as she exclaimed, "I *love* pomegranates!" I do too, Martha.)

1 tbsp sea salt dissolved in 2 c water

1 tsp dried Greek oregano or other dried herbs

Juice of half a lemon (or lime), lemon or lime rinds cut into slivers (to look pretty!)

2 cloves garlic

Olive oil: pour about ¼ to ½ of an inch over the top of the marinade

Instructions

The olives will be ready to eat after sitting in the marinade for just a few days. Store, still in the marinade, in a cool pantry, away from light. If kept too long, the lemon and vinegar flavors will predominate, so eat these within a month after they are ready. *Or*, pour off ½ the marinade, use as salad dressing, and fill remainder of jar with Global Gardens Balsamic Fruit Vinegar to store in canning jars.

YOU DON'T ALWAYS NEED THE OIL

Using vinegar as the sole condiment on salads, steamed and even raw veggies is not only tasty, it's healthy and culinary chic too! Try cooking with vinegar in a wok or skillet without any oil. Even ice cubes made with vinegar make a terrific cocktail (see page 105). I'll admit, as a kid and prior to making my own vinegars, I thought I didn't like vinegar. Plenty of people come into the Global Gardens Destination Store and tell me just that. Then I pour them a swig. Seeing the enjoyment on their faces, hearing them ask for another sample, this time a different fruit, I know I have a convert on my hands . . . for life!

CREDIT: ROBERT DICKEY

GLOBAL GARDENS BALSAMIC FRUIT VINEGARS USED IN THIS BOOK, RECIPE PAGE REFERENCES

I'm always dreaming up new balsamic fruit vinegar formulas to pair with California Golden or Dark Balsamic. Shown here are some favorites profiled in this book. As of this writing, I have developed a terrific Pear Ginger, Black Currant, and Berry Cranberry. I work with in-season fruit and different flavors, so we don't always have every flavor year-round. Most of the vinegar recipes in the book can be made flavorfully by swapping your favorite fruit flavor for mine. I look forward to hearing your ideas!

1. **Apple Ginger Golden (**Recipe pages 105 and 186)
2. **Black Currant Golden (**Recipe page 70)
3. **Blood Orange Dark (**Recipe pages 112, 177, and 224)
4. **Fig Dark (**Recipe pages 57, 61, 96, 118, 169, 170, 191, 193, 195, 196, and 200)

5. **Mango Golden (**Recipe pages 105 and 120)
6. **Meyer Lemon Golden Bliss (**Recipe pages 72, 108, 115, 126, 152, and 163)
7. **Pear & Tamarind Golden**
8. **Passion Fruit Golden (**Recipe pages 105, 147, and 182)
9. **Pomegranate Golden (**Recipe pages 105, 145, 148, 166, 210, and 220)
10. **Raspberry Dark (**Recipe pages 76, 87, and 105)
11. **Strawberry Golden (**Recipe pages 58, 99, 105, and 129)

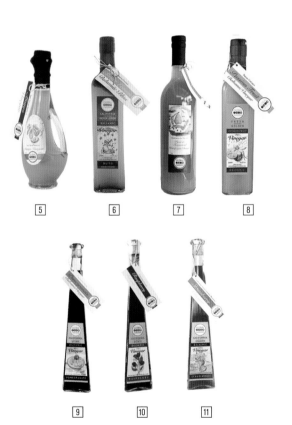

CALITERRANEAN™ BALSAMIC RECIPES

REDUCTIONS MADE SIMPLY AND TASTEFULLY

Have you ever seen those fancy loop-the-loops on fancy restaurant plates, tasted the trail, and wondered what the goodness was all about? Typically, it's a reduction of something. At my house, it's a reduction of Global Gardens Balsamic Fruit Vinegar. I use a lot of vinegar reductions in this cookbook— sometimes I add things to them (see Orange Fantasia Cake, page 224), and sometimes they are very straightforward.

The Global Gardens Olive Oil and Gourmet Food Club includes one bottle of our latest balsamic fruit vinegar in each season's shipment (delivered every three months). That might sound like a lot of vinegar, but not if you're cooking even one out of every three days. Getting four bottles of vinegars in varying sizes per year encourages my customers to try making reductions.

I'll admit, *all of us* have ruined one reduction and overcooked it, making thick burnt vinegar caramel, or worse, hard candy. Do it once and you'll never repeat your mistake. Global Gardens Balsamic Fruit Vinegars reduce easily because they contain a high ratio of fruit. The natural sugars in the fruit thicken and become syrupy as the acid from the vinegar is being boiled away. Other vinegars might reduce, but you'll need to add something like brown sugar and sometimes even butter.

Method

One cup Global Gardens Balsamic Fruit Vinegar will make ½ cup reduction. Pour into open skillet. You can use a saucepan; however, it will take longer for the acid to cook off—the larger the skillet, the faster the reduction will occur. Engage your exhaust fan; boiling vinegar will sting your eyes, so don't lean over the skillet or breathe the steam. It certainly won't kill you, but it won't feel too good either. Turn the heat to medium high and bring vinegar to a boil. When the reduction begins, the vinegar will get very foamy. Pay attention because this is "make it right or make hard candy" time! When the color of the vinegar (not the foam) sticks to the back of a metal spoon, it's finished. You can also note a line on the skillet when you poured the vinegar in and make an imaginary line at the halfway mark to show you when to stop cooking the vinegar. It will seem too runny but will thicken as it cools. Reduced Global Gardens Balsamic Fruit Vinegars will keep rather indefinitely in the refrigerator. I like to store them in a plastic squeeze bottle so I can use them spontaneously to make tasty squiggles or even write messages on plates. See the *MV* initials topping the Mar Vista Crème Brûlée (page 220)—the Pomegranate Balsamic reduction completely makes this dessert zing with ultimate flavors.

A lovely way to store and use balsamic reductions is in our Global Gardens

exclusive vinegar reduction bottles. Bandhu Dunham, owner of Salusa Glassworks in Prescott, Arizona, hand-blows each bottle, giving it a distinctive swirl with a one-of-a-kind, original finish. Now here is a guy living the Caliterranean lifestyle in every way—made in the USA, healthy, and artisanal!

Here is a partial list of things made delicious by Global Gardens Balsamic Vinegar reductions. Reductions are tart and add a wonderful balance to rich and savory things like:

Brie cheese
Chèvre cheese
Cream cheese
Salmon
Steaks
Chicken
Pork chops
Pasta dishes (use in place of oil! See Seaweed Power Pasta, page 169)
Chocolate brownies
Cheese tarts
Cakes

SALAD DRESSINGS PERFECTED

Have you read the ingredients on a bottle of salad dressing in the grocery store: dextrins, high fructose corn syrup, MSG.? Some of them might even taste good, but I'm with Michael Pollan on this one. If you can't read the name of an ingredient, it's probably bad for you, and you don't want or need them to make a salad dressing memorable. Generally, you can make an excellent dressing using a 3:1 ratio of extra virgin olive oil to vinegar . . . works every time! Always use finely ground sea salt and fresh ground pepper to bring out the flavors of whatever you're dressing.

Speaking of memorable, Brad Linderman from the Greenwood Pier Inn, Elk, California, was kind enough to share his famous salad dressing with me for this book. Elk (originally known to some as Greenwood) is a picturesque town of only about two hundred people, situated directly on a super scenic section of Highway 1, just south of Highway 128. The Greenwood Pier is one of my favorite B and Bs in California, for its dramatic views and cozy, eclectic rooms. Perched on a cliff, with stunning lookout points in every direction, it's one of those places you can't describe even when it's right in front of you. It's something you need to feel. Don't miss Bowling Ball Beach just a few miles away!

YIELD: ABOUT 2 CUPS

Ingredients

2 c packed broadleaf basil, de-stemmed

2 tbsp organic sugar

1 tbsp white pepper

1 tbsp finely ground sea salt

¼ c Dijon mustard

¼ c water

2 tbsp Global Gardens Fig Balsamic Vinegar

1 c Global Gardens Ascolano Extra Virgin Olive Oil

Method

Blend all ingredients together in a blender or food processor. Use on salads, roasted or grilled vegetables, and seafood. Don't refrigerate leftovers. Store them in the pantry in a tightly sealed jar. Good for about one week after making. Sit back, enjoy the view, and drink in the green goodness that is the Caliterranean life!

IDEAL VINAIGRETTE

A bit of controversy swirls with foodies and homemade salad dressing enthusiasts about vinaigrettes. What's the difference between a vinaigrette and a plain ole salad dressing? For me, that's an easy answer. Switch the ratios, 3:1 vinegar to olive oil. Vinaigrettes are awesome low-fat, high-flavor condiments. Tangy, yes, but you'll love it too as the perfect toss for a nicoise-style poached salmon salad, tuna salad, lightly steamed asparagus, broccoli, or, really, any lightly steamed, crunchy veggie. Try oven roasting corn on the cob by running each husked ear with 1 tsp Global Gardens Kalamata Extra Virgin Olive Oil. Sprinkle generously with salt and roast at 350 for 20 minutes. Cut the corn off the cob and toss in freshly diced cucumber, scallions, and halved cherry tomatoes. Add a boiled egg for protein and sprinkle generously with a Pomegranate, Fig, or Mango Vinaigrette. *So flavorful and good for you*, so Caliterranean!

MEMORABLE MARINADES

I was not born with a proverbial silver spoon in my mouth, but I was born to love to cook and eat. I can't imagine being challenged by things like boneless, skinless chicken breasts, or lean center cut pork chops. But time and time again, I've had terrible examples of each—at nice restaurants and at the homes of some nice friends. Marinating lean meats and seafood ensures better flavor with little work, not even what I'd call planning. Many times I'll leave for the day, look in my fridge, and find an empty cavern that tells me to pull something out of the freezer for dinner. Out come the Ziploc bags, extra virgin olive oil, and balsamic fruit vinegar. Any kind of meat will benefit from thawing in a terrific marinade, and it will be ready to cook when you get home. Seafood, on the other hand, should not be marinated for more than 15 or 20 minutes . . . and never in a vinegar-based marinade or it will start to cook! (See Scallops Ceviche with Red Curry Delight Sauce, page 145.)

The best simple marinade for any kind of meat will be a ratio of 2:1, vinegar to olive oil. Don't forget to add sea salt generously. Remove frozen meat from its packaging, running cold water over it to peel away any remaining paper or plastic. Pull apart into individual pieces to fit into a Ziploc bag containing your marinade. Place in refrigerator, square side down, not standing, to allow the marinade to permeate the entire package, not just the bottom. Add a sprig of rosemary, tarragon, oregano, or other fresh herb to the top of each serving of meat, allowing the marinade to soak up from the bottom of the bag. If you like garlic, add a minced clove. Grilling, oven roasting, or skillet cooking is just minutes away when you get home. If you don't feel like cooking, the marinade will hold the meat until the next day. Seafood marinades are best made with olive oil, spices, and fresh herbs only.

CALITERRANEAN

WHAT DOES CALITERRANEAN MEAN?

Who came up with the word *Caliterranean*? I did! Or . . . I *think* I did. Several years ago I was in the shower, thinking about being raised as a Greek American, the food, the livelihood, my subsequent adventures to Greece . . . then my mind kind of drifted to what I do now: I am an olive farmer, a food purveyor, photographer, graphic designer, and lover of feeling good, exercising both mind and body, creating new flavor dynamics, entertaining, learning from the experiences of my immediate environment (whatever that may be, wherever I am), and appreciating my life. (Yes, I take long showers!) I started to think about how, since I've become a specialty food producer, I have gotten a much better grip on my weight and understanding of how what we eat affects our metabolism and well-being. *It's so much more than the Mediterranean diet*, I mused.

I'd felt an urge to see California even as a young girl. (I'd threatened my mom even when I was little, "I'm going to California!" in response to her reprimand of the moment. "What do you know about California?" she would tease accordingly. Finally, when I was twenty-five, I answered the call and went for the first time—for a life full of sunshine, surf, bronze tones, fresh seafood, sailing, smiling, happy people—oh yeah, I'd idealized this place—or had I? Corny as this all sounds, it's true, and the word *Caliterranean* just washed over me like the feeling of completing a huge project.

I stood for a while and thought about my new word. Is it difficult to say? Well, not really any different than Mediterranean might be. Is it a brand? Or more contemplatively, is Caliterranean a lifestyle, a diet, a garden, a grower, a student of life? The answer has developed into a brand that will encompass all of these aspects significant to my past development as a designer/writer, then farmer, food producer, and seeker of personal and professional relationships with others of like principles. At this point, the hot water heater literally couldn't take it anymore, ceasing to produce the heat necessary to continue my deliberations.

I googled the word *Caliterranean* immediately—nothing came up. I asked some friends and family members what they thought. I toyed with it for about a week as the concept of what Caliterranean could become solidified. I endeavor to grow the Caliterranean brand further— into a flavorful, substantive Caliterranean diet that will help people achieve their weight goals, more examples and how-tos regarding Caliterranean living, creating a movement called Caliterranean growers, and even transforming my home into the first of many Caliterranean homes (after all, the hot water heater in my home should be solar operated so it will never run out of hot water, right?). The exciting thing to me, regarding all of this, is that anybody, anywhere, can eat and live the Caliterranean way.

THE HEALTHY CALITERRANEAN LIFESTYLE

You'll discover the huge number of healthy benefits from cooking with extra virgin olive oil and vinegar throughout this book. There are even studies being done as a follow-up to a 2005 article from *Nature,* which revealed that the antioxidant oleocanthal (found only in authentic extra virgin olive oil) has an effect similar to ibuprofen in that it inhibits an enzyme (cyclooxygenase) that causes inflammation and pain. Most test patients only required two daily teaspoons of extra virgin olive oil to see a notable improvement in chronic symptoms.

Living the healthy Caliterranean lifestyle includes taking control of persistent issues that inhibit us from realizing the type of life we want to experience. It means creating a more natural abundance in our immediate environments with less dependence on foods and objects imbibed with chemicals. It means educating ourselves on GMOs and our local food chain. Hormones shot into dairy cattle to yield more milk or into animals to create processed meats are detrimental to our overall well-being, diluting the very reason we think we are ingesting them. Aren't we all too busy to know these things, read about them, protect ourselves and our families? *We don't have the time not to.* Regulations on food and labeling change so frequently, there is no practical way to keep track of it all, but there are a few simple steps we can take at home, which, cumulatively, will make a huge difference. One of the best ways to learn is to get into a conversation with your local/regional growers at your nearest farmers' market or food courts, events, and festivals that feature local, homemade vendors.

Know where your food comes from. Whenever possible, eat local! Most states have their own dairies and farmers. Eat food that is in season; there is no reason to buy frozen vegetables grown in China. Locavore movements are all over the place, and even if you live in an area with no farmers, within your own state there are certainly food cooperatives or local producers of *something*. The general desire for a reduction in the carbon footprint on planet Earth is part of the spawning of eco-consciousness and sustainability—all components of the Caliterranean lifestyle.

Have you visited ediblecommunities. com? From *Edible Boston* to *Edible Santa Barbara* and, currently, eighty other regions of the United States (plus Toronto and Vancouver), this magazine won the James Beard Foundation's 2011 Award for Publication of the Year. But it's not just about food publication awards when it comes to the Edible Communities organization. Sincere strength in food news reporting

and informational services makes Edible publications a standout, connecting consumers with regional farmers, family growers, chefs, and food artisans who share the economic viabilities of their own communities as individuals and professionals. Printed Edibles publications are typically free of charge (if you can get to them soon after they are published) and printed using soy-based inks on lovely recycled paper. Guaranteed subscription delivery is available (and worth it!). The photography is absolutely delicious. Easy to comprehend, with generous information that continually changes, keeping up local agricultural issues, including recipes and other locavore information, it allows us to travel virtually to the other destinations' food resources, preparing us for vacations in other parts of North America, and giving us years' worth of fantastic foodie reading!

Where can we start anyway? First, in our homes, then in our schools. Food Corps (food-corps.org) is an organization to follow (and participate in). The vision of this nonprofit is for AmeriCorps service members to build and tend school gardens, developing farm-to-school programs around the country. There are a lot of ways to get involved in this movement; educating yourself comes first! Their website is comprehensive, friendly, and updated weekly.

The FDA cannot police every product label (and they don't, especially on extra virgin olive oil), let alone every edible plant that goes from some farm somewhere onto our dishes and into our mouths. The education has to begin somewhere. The easiest place is in your own hometown, but the second easiest is on the Internet. Never before have so many good food resources been available to us. This is a short list to get you started in your home.

Learn the recipe for meditation. I remember when I learned to meditate back in 1995. I was at the height of vocational stress, gaining weight, feeling tired all the time, and, at thirty-five, too young to feel so hopeless about everything. Meditation was, and still is, the only way I can clear my mind and start all over again. It's another one of those things you don't have time not to do. *Meditation gives you more time in your day, I promise.* I have so many people ask me how I have so much energy to get done what I do in a day, how I maintain my weight, how I appear (key phrase) to be so relaxed and in control.

There are so many easy ways to quiet yourself for just a few minutes a day. You can join some fancy class or create a meditation group and make it a more powerful experience cumulatively. Whatever you do, just try it. Give it a full week, twenty minutes per day. If you think there is absolutely no possible way to extract twenty minutes from your overloaded, maxed-out life, do it anyway. Wake up twenty minutes earlier or go to bed twenty minutes later. The very best time for me is right in the middle of the afternoon.

Try this simple, effective recipe for meditation: Close your office door and turn off all electronics that can bong, bleep, or bling. Close your eyes, sitting up straight in a comfortable chair with both feet on

Theo's meditation room

the ground. Concentrate only on your breathing . . . in, then out. Little voices will remind you of things immediately—the rip in your son's baseball pants or the fact that your daughter needs a ride at three instead of four today. Don't force yourself to stop the voices, but as soon as you realize that they are dominating thoughts of your breathing in and out, go back to that simple concentration of breathing in and out. Don't set any alarms that will jar you from your meditation. Instead, open one closed eye and look at your watch or other clock nearby and check the time. Within a few days you will easily get to twenty minutes without looking. Sometimes you might even fall asleep for a few minutes. That's OK. Don't judge yourself; you body needs it. Statistically, if you do something for seven days straight, it becomes a habit.

If you stop for just three days, you lose the routine. Over 1,500 studies since 1930 exist regarding measurable attributes of meditation. The short list includes a decrease in blood pressure, anxiety, insomnia, and the risk of disease in general.

Exercise your body. This is nothing new. You know it, everybody knows it. The challenge, if you're not doing it already, is to find something you love—whether it's a brisk twenty-minute walk at lunchtime, a spinning class, Pilates, yoga, or a neighborhood softball team—the truth lies in just beginning to do it, the rest will follow.

The healthy Caliterranean lifestyle relies on a very simple, effective, life-changing combination of eating right and being good to your mind and body.

CALITERRANEAN COOKING

Join our e-mail list at caliterranean.com if you're interested in a comprehensive guide to finding your individual body weight through nourishing your body easily and effectively, without giving up things you love like pastas and bread. *Olive Oil and Vinegar for Life* shares the beginnings of the Caliterranean diet by giving you great recipes using the healthy fat of extra virgin olive oil, along with some great flavors and healthy attributes using Balsamic Fruit from Global Gardens. I am working on a comprehensive Caliterranean Diet, which will feature research-based knowledge from nutritionists and MDs who will substantiate my program, offering advice, recipes, and detailed formulas for a healthier, tastier life; stay tuned!

Caliterranean cooking is easy when you use fresh ingredients like the ones featured in these recipes. The difference between fresh versus frozen is unparalleled. If you need frozen foods for convenience, buy fresh and then freeze it yourself. Berries readily available all summer long will freeze beautifully, *unwashed*, in just a few hours on cookie sheets, sliding easily into freezer bags for happy winter thawing, washing, and eating. Super easy tips on what to freeze, easy preserving, drying, and canning for the entire family can be found in my favorite food book originating in 1971,

Putting Food By, now in its current fifth edition by Janet Greene, Ruth Hertzberg, and Beatrice Vaughan.

Whatever you do, only buy range-fed and organic chicken meat. Beef and pork should be grown without hormones or antibiotics. Visit www.seafoodwatch.org for recommendations on purchasing seafood. Farmed seafood, especially salmon, is often fed with cat food or corn pellets to fatten the fish, destroying valuable healthful qualities and flavor. Try to buy your meats and seafood from a source you or your local meat man knows to be a reputable producer. The flavor and texture of "real" meats versus the highly processed quantities that are cranking into the big box stores offers unparalleled sweetness you and your family deserve.

Reaching beyond local food purveyors to resources around the country where only certain nutritional foods are available is great when you know the origin. One of my favorite things to do when I'm on vacation or traveling on business is to visit the local farmers' market or artisanal food center. Communities large and small have them. Hey, we all have to eat, right? Eating out all the time begins to taste bland and processed. (Do you realize how many sauces, soups, and condiments your favorite restaurants call "homemade" come in five-gallon buckets?) It's so advantageous to take some fresh things home, have some great extra virgin olive oil or balsamic vinegar on hand, and whip up dinner . . . in a lot less time than you think! There is a false perception out there that cooking is difficult, that it doesn't taste as good when you make it, that it's easier to just go pick something up (or have it delivered), and that, overall, cooking is just a royal pain. *False*! Just a quick conversation with artisanal food purveyors at any local market around the world will open you up to easy and endless possibilities.

Donna Bishop is just one of those people I met at a small farmers' market up in Gualala, California, which is situated in the banana belt of Mendocino County. It was love at first taste of Donna's seaweed snacks, seaweed salt, and her seaweed conversation! One of the world's largest concentrations of edible seaweed is on the Mendocino Coast, where Donna harvests in compliance with all California Seaweed Harvesting Regulations. It's a dangerous and tedious task, and she takes extreme care in cleaning, drying, processing, and packaging to produce and sell the highest quality and grade of seaweed products. Seaweeds have been used around the world to cure and prevent diseases, with major research currently under way for seaweed extracts in cures for cancer, AIDS, and diabetes, to name a few. Kombu has high concentrations of iodine 127, which, by ingesting only three to five grams a day, will pull radiation and other heavy metals out of the human body. All edible seaweed varietals are higher in vitamins and trace minerals than any other land vegetable. See my Seaweed Power Pasta recipe on page 169, inspired by Donna Bishop and part of a twelve-course meal Daniel and I made for twelve people (including Donna) at Mar Vista Cottages in Gualala. Donna's products can be ordered from gualalaseaweedproducts.com. (See more information on Mar Vista Cottages, page 166.)

Lora La Mar and her husband Bob are two more dear examples of the rewards of being inquisitive at "other" local farmers' markets. Mendocino Sea Salt is the result of seawater collections carefully selected over deep undersea canyons known for producing mineral-rich upwellings. The seawater is simmered until delicate crystals of salt begin to appear on the surface, growing and sinking to form a salt bed. Bob and Lora's salt is distinctively light because of careful monitoring during the following heat and evaporation process, resulting in a snowflakey crispness that is not coarse, yet full of flavor. See recipes on pages 123, 124, 141, 157, 166, and 195 featuring this handcrafted wonder. Their Mendocino Sea Smoke is produced by carefully smoking the handcrafted white salt crystals in a specially built smokehouse, an unforgettable rich, smoky color and flavor used in the recipes on pages 170 and 219.

One of the biggest rules in Caliterranean cooking is: no sugar! Seldom is actual sugar necessary in preparing meals. Talk to any chef and, if he or she is honest, they will divulge that sugar and butter are their two biggest secret ingredients. No wonder going out to eat makes us crave more carbohydrates just a couple of hours after we are "stuffed." Refined sugar is why. Eliminating it altogether, or using raw organic sugar when necessary, is the answer.

CREDIT: LORA LA MAR

Edible flowers and wild greens abound all over the country in front yards and planted garden beds purposefully designed as edible flower gardens. The greatest example I saw was at the Farmhouse Inn in Forestville, California, family-owned and operated by Catherine Bartolemei and her brother Joe. These fifth-generation Russian River Valley vineyard owners are dedicated to the preservation and promotion of Sonoma County's diverse agricultural offerings while simultaneously offering warm hospitality, delicious food, fine wine, a pampering spa, and cozy accommodations. Over thirty edible flowers grace the Farmhouse Inn gardens, many I had never tasted or seen before. Edible flowers are not only beautiful on the plate, but add a natural spicy or savory eclipse to any dish. These wild greens can include nettles (yes, nettles), clover, mustard, miner's lettuce, dandelions, and purslane. These edible plants add dimension to our cooking—not to mention high omegas and dietary iron too.

Culinary lavender flows freely just two properties away from mine in Los Olivos, California. Clairmont Farms produces organic lavender products the most wholesome way. One of my favorite reasons to visit my neighbors is their driveway—planted with olive trees by Spanish missionaries over 180 years ago. Hand-harvested lavender is sun-dried, thrashed, sifted, and steam-distilled. Clairmont's lavender is sweeter because it has been distilled with a low-temperature process in copper. Find out more by

visiting their website at clairmontfarms. com. Used for millennia, lavender is the best-known herb for its medicinal properties, but Clairmont also grows and packages an excellent culinary lavender. Caliterranean cooking would never be complete without herbs. You can grow them easily in all fifty states, even in small pots—most of them grow like weeds, with only water necessary to nourish them, and what a difference they make to the most simple of dishes! (See page 224 for my Orange Fantasia Cake recipe with a surprise flavor of culinary lavender used for the glaze.)

Yes, I always use organic—everything I can get my hands on—a Caliterranean must. Why? For flavor, quality, and the simple fact that we do not need chemicals to grow food. Nonorganic pesticides are ruining our ecosystems by destroying many species of plants and animals as well as our water supply. I always support my local farmers who claim to be pesticide free— you can tell they're being honest because their fruits and veggies aren't perfect . . . like nature intended. It's really tough for the small farmer to afford government- contracted third-party organic certification agencies. But do go ahead and taste-test a nonorganic fruit or veggie side by side with an organic one of the same varietal. Taste the difference in the organic sweetness versus the bitter or flat flavor that prevails in nonorganics.

Caliterranean cooking means experimenting in all phases of a meal, from the beginning to the end. The key to success is making the process low stress. Use fresh extra virgin olive oil as a necessary monounsaturated fat. It will enhance the flavor of anything you are cooking or baking, as will a high-quality balsamic fruit vinegar. Forage, play, pretend you are an artist creating a palette (and palate) full of colors and experiences. There is definitely an art to this . . . but it's not an art that can be judged or mistaken. It's an art of using natural ingredients and nature's own bounty from our backyards, local farmers, and even mail order from food artisans around the country. The goal of Caliterranean cooking is to increase our awareness of what tastes good, what ingredients are available, and the involvement of that knowledge into the creation of flavors we and our loved ones will surprisingly enjoy.

Finally, Caliterranean cooking promotes a new kind of sustainability with a conscientiousness to reducing waste. You will make your carbon footprint a lot healthier by going to the grocery store or market no more than twice a week.

Make a list. Don't buy things because they look good and do so only to cross something *off* your list. When the end of the week rolls around, you don't want biological experiments growing inside your refrigerator. The few minutes it takes you to organize a list of necessary recipe items and meal planning for the week will take a lot of guessing and frustration out of meal preparation time. Caliterranean cooking should be fun, fresh, mindful, tasteful, simple, and memorable!

GLOBAL
GARDENS
FEATURED
PRODUCTS

Global Gardens Extra Virgin Olive Oils, Flavor
Profiles, Suggested Uses, and Recipe Page
References   48

GLOBAL GARDENS EXTRA VIRGIN OLIVE OILS, FLAVOR PROFILES, SUGGESTED USES, AND RECIPE PAGE REFERENCES

Following is a list, with photos and flavor profiles, of just a few of Global Gardens' featured extra virgin olive oils in this book.

Balsamic fruit-infused vinegars are on page 26. Keep in mind that sometimes we change the bottle shapes, colors, and even the label designs, so if you can't find what you're looking for,
please e-mail me personally at theo@globalgardensonline.com or info@globalgardensonline.com.

Global Gardens either grows, harvests, or bottles each of the olive oil fruit varietals shown here in alphabetical order. I have chosen them because of their own individual flavor profiles and specify them by varietal in recipes to achieve maximum character in the finished dish. Subjectivity prevails! Please experiment with your own favorite varietals; an herbaceous finish is as perfect a balance to desserts as it is to pastas and salad dressings. The only thing I will caution you against is in not using the more robust varietals when pairing with chocolate or sweet desserts—personally, I don't think it works because the culinary bitterness is not symbiotic with sugary flavors, throwing them off balance.

If you prefer to use your own favorite extra virgin olive oils in the recipes herein, simply follow the flavor profile referenced for each oil, matching that flavor nuance with the recipe itself. I am hoping to hear from my readers with your own favorite pairings and experiments with worldly flavors à la Global Gardens!

Ascolano: Buttery, fruity, mango/banana tropical finish. Great replacement for butter, desserts, salad dressings. Recipes on pages 76, 80, 92, 94, 99, 119, 124, 185, 186, 213, 216, and 219.

Frantoio: Lush and fruity with a medium white peppercorn finish. Lovely for breakfast dishes and a perfect all-occasion cooking or salad oil.

Blood Orange: Fruity, definitive orange flavor. Baking with desired orange flavor, citrusy marinades, fruit salads, pasta tosses with fruits and nuts. Recipes on pages 112 and 224.

Kalamata: Fruity, buttery. Perfect for phyllo recipes, frying, salad dressings, pastas, butter replacement. Recipes on pages 58, 61, 67, 120, 135, 163, 208, 210, 215, 223, and 226.

Koroneiki: Buttery, mildly peppery finish but sweet simultaneously. Great for baking, phyllo recipes, frying. Use as a natural skin moisturizer. Removes mascara easily; terrific as eye conditioner and wrinkle prohibitor for many people. Men can shave with it. High polyphenols in this varietal make for a great skin product! Recipes on pages 82, 92, 120, 124, 126, 129, 141, 208, 215, 226, 231, and 233.

Farga: Robust, herbaceous, peppery finish. Salad dressings, marinades, pastas. Cheeses. (See green label in our 10th anniversary bottling shot.) Recipes on pages 61, 64, 72, 148, 152, 164, 174, 178, 180, and 185.

Los Alamos Signature Estate Blend: Robust with a salty, rosemary finish. Reveals hidden flavors from pastas and bigger dishes.

California Koroneiki: The same varietal as the Greek Koroneiki, but grown at our Los Olivos Farm Stand. Super fruity with essences of green apples and a woodsy finish of pears. Perfect for salad dressings, marinades, and oven roasting. Can be substituted for Manzanilla in recipes shown on pages 135, 193, 195, 199, 200, and 202.

Meyer Lemon: Meyer Lemon rinds crushed simultaneously on the mill with fruity black Mission olives. This is not an *infused* oil! Lovely for salad dressings, seafood, pastas, lemon flavor desired in baked goods and protein dishes. Surprisingly versatile. Recipes on pages 92, 96, 107, 123, 124, 131, and 208.

Mission: Attention! Mission Extra Virgin Olive Oils may be mild, medium, or robust in character. We bottle buttery, and when we have a good yield we'll harvest early to also get a robust version. Our crops differ from year to year, and the website is always up to date. Recipes on pages 111, 115, 132, 169, 170, 172, 215, and 226.

California Estate Mission: Super buttery, great for vinegar blends, marinades, baking.

Mission/Manzanilla Blend: Fruity with a kicky, sagey, grassy finish. Amazing for bringing out the flavor in any dish, especially vegetarian. We call it Kiss My Grass! Recipes on pages 78, 84, 118, 138, 147, 199, and 215.

Stallion Grove Mission: Robust. Lush and fruity, followed by a grassy, green peppercorn finish. Amazing for marinades, pastas, fresh stand-alone olive oil. You can count on our website to be current and informative.

Santa Ynez Italian Varietal Blend: Medium fruitiness with bright pink peppercorn finish. Perfect for all—occasion use. Recipes on pages 91, 100, 166, and 182.

AMAZING AIOLI

Love mayonnaise? I sure do. I love it on typical things like hamburgers and leftover Thanksgiving turkey sandwiches, but I really love it on french fries, seafood of all types, and as a fresh veggie dip. Aioli is a simple, healthier way to get the taste of mayonnaise without all the "stuff" that's on the label of your commercial brand. (Seriously, look at the ingredients, they're scary!) The only thing about aioli is that you should eat it right away. Raw egg products don't keep very well. Make this when you can share it. It's wonderful for freshly steamed artichokes too!

Ingredients

1 egg

1 garlic clove, crushed

1 tsp Global Gardens Indian Curry mustard

1 tbsp finely chopped chervil (or basil)

½ c Global Gardens Extra Virgin Olive Oil (any varietal will work; choose the flavor profile you like best!)

Pinch of salt and 3–4 grinds of black pepper

Method

In a food processor, blend egg, garlic, mustard, and herbs. Very gradually add the olive oil using the drip pipe. When it has thickened to the texture you like, season with salt and pepper and use immediately.

CELEBRITY CHEF FEATURES

I thought it would be fun, informative, and special to ask a few celebrity chefs to celebrate the debut of Caliterranean cuisine by providing some of their favorite recipes using olive oil and vinegar. The response was overwhelmingly positive, the flavors cleanly sublime, and the experiences each truly original and inspiring. My own cooking and palate matured immensely with these four very different, vibrant chefs, adapting and creating new recipes themselves for

Olive Oil and Vinegar for Life. I am very appreciative of being introduced to them all by the Santa Barbara Convention and Visitor's Bureau (santabarbaraca.com) and the California Travel and Tourism Commission (visitcalifornia.com). And I am grateful for finding myself here in California! Each recipe in this "Caliterranean celebrity chefs" section was created and written by the chef himself.

CREDIT: JEREMY BALL

CHEF BRADLEY OGDEN

Root 246 • 420 Alisal Road • Solvang, California • 805.686.8681 • root-246.com

Chef Ogden, culinary genius behind over fifteen California and Nevada restaurants and named "Best Chef of California" by the prestigious James Beard Foundation, is perhaps best known by foodies nationally for his restaurant at Caesars Palace, Las Vegas. Chef Ogden originated a restaurant in the heart of Santa Barbara, California Wine Country, Root 246 in Solvang. He has since moved on to northern California, creating and continuing his commitment to American cuisine that is farm fresh, relatively simple, and absolutely delicious.

One of the things I enormously love about eating at Bradley's restaurants is that I can actually taste every ingredient he uses in his recipes. Nothing is masked with heavy butter or mysterious sauces. A tour of his kitchen reveals an inspiring assemblage of professional exception. Every day, something "else" is happening relative to producing year-round California flavors in Bradley's cuisine. They have their own dehydrators, vacuum packers, canning equipment, and every culinary talent known to man. Inspiringly efficient, it is really phenomenal to see each daily produce delivery individually laid out on trays, layered in a baking rack, for inspection. Every item is turned over to ensure freshness, color, and presentation. One can truly taste the Caliterranean freshness in Bradley's cuisine.

Thank you, Bradley, for sharing two of my favorite recipes from your restaurant, Olive Oil Poached Arctic Char (on that killer bed of color and flavor!) and your famous Caesar Salad. I have tried the char recipe on salmon, halibut, and scallops. The Caesar Salad is amazing with all Global Gardens Balsamic Fruit Vinegars.

CREDIT: JEREMY BALL

OLIVE OIL POACHED ARCTIC CHAR WITH BEETS AND BLOOD ORANGE

SERVES 4

Ingredients

4 fillets of arctic char, 5 oz each with skin on

2 c Global Gardens Extra Virgin Italian Varietal Blend Olive Oil

1 c white wine

Lemon and orange peels

Bay leaf

Dill sprigs

Kosher salt

18 fresh baby beets (red, gold, chioggia or candy cane, 6 each)

½ c Global Gardens Fig Balsamic Vinegar

1–2 blood oranges

Procedure

Buy arctic char steaks from fish market or prepare yourself. Steaks should be 1-inch thick with center bone and pin bones removed. Lightly warm ingredients, excluding arctic char, and steep together for 15 minutes. Gently poach fillets, lightly simmering in liquid for approximately 10 minutes.

Procedure

Wash beets well, leaving 1–2 inches of stem attached to prevent loss of color and nutrients during the cooking process. Cook each variety of beets in a separate pot. Cover beets with cold water and a dash of salt. Simmer beets lightly for about 10–15 minutes until cooked but still firm. Drain and place on a cookie sheet; let cool. Gently rub skins off with a clean kitchen towel. Cut beets lengthwise in halves and reserve for service.

Beet Reduction

Add beet juice and balsamic vinegar in a small noncorrosive saucepan, reduce by ⅔ or until it lightly coats the back of a spoon. Strain beet reduction from pan and reserve for service.

Plating

Place the beet reduction at bottom of bowl.

Place assorted beets over reduction.

Place arctic char over beets.

Top with segments of fresh blood orange.

Garnish with pea shoots or microgreens for color.

CREDIT (OPPOSITE PAGE): JEREMY BALL

BRADLEY'S CAESAR SALAD

SERVES 4

Ingredients

2 heads romaine lettuce

2 large garlic cloves, minced

½ tsp capers, rinsed and minced

6 anchovy fillets, mashed
   with a fork

2 egg yolks

¼ tsp dry mustard

2 tbsp Global Gardens Strawberry
   Balsamic Vinegar

¼ tsp kosher salt

¾ tsp cracked black pepper

½ c Global Gardens Kalamata Extra
   Virgin Olive Oil

Parmesan Croutons (*see recipe below*)

½ c shaved Parmesan

Procedure

Trim off any brown or bruised leaves from the romaine lettuce. Tear the leaves into 2-inch pieces. Wash, dry, and refrigerate.

Combine the garlic, capers, and anchovies, mixing together to form a paste. Add the egg yolks, dry mustard, vinegar, salt, and ¼ tsp cracked black pepper. Whisking continuously, very slowly add the ½ cup of olive oil. Continue whisking until all the oil has been added and the dressing is thick and smooth like mayonnaise. Refrigerate the dressing for 30 minutes to develop its full flavor.

Place the romaine in a large bowl and add the remaining cracked black pepper. Pour the dressing down the sides of the bowl, lifting the lettuce up and over, coating the leaves evenly. Add the croutons and toss. Place on plates and garnish with shaved Parmesan.

Egg Poaching Technique

1 qt cold water

2 tsp white vinegar

1 tsp kosher salt

In a 2-quart stainless steel wide pot, combine ingredients and gently bring to a simmer.

Crack egg in small dish to pour down the side edge of the pan. Poach until set (3–5 min).

Remove from pan with a slotted spoon and place on salad.

Parmesan Croutons

6 small cloves garlic, peeled and crushed

¼ c Global Gardens Kalamata Extra Virgin Olive Oil

2 c French bread, cut into ¾-inch cubes

½ c grated Parmesan cheese

Procedure

Preheat oven to 350°.

Combine the garlic and 1 tbsp of the extra virgin olive oil in a small saucepan.

Place over moderate heat until the garlic has browned. Remove from the heat and discard the garlic cloves.

In a bowl, toss the bread and remaining olive oil, evenly coating the cubes. Place the bread cubes on a sheet pan and bake for 15 minutes. Stir them 2 or 3 times while baking. Once the croutons have become a deep golden brown and are crisp all the way through, remove them from the oven and place them in a large bowl. Add the Parmesan cheese to the croutons while they are still warm and toss the croutons and cheese together.

CHEF PAUL MCCABE

L'Auberge Del Mar • 1540 Camino Del Mar
• Del Mar, California • 858.259.1515 •
laubergedelmar.com

Kitchen 1540 at the L'Auberge Del Mar hotel blew me away. Now that's a typical California surfer thing to say, but I'm not sure how else to explain the flavor surprises Chef Paul McCabe incessantly tantalized me with, along with the cozy Caliterranean environment offered on the entire property. Nestled in the heart of downtown Del Mar, right on the seemingly endless white sand beach, L'Auberge Del Mar welcomes the traveler with a private path to the beach, plush accommodations, a sumptuous spa, and, most of all (for me), refreshingly original dining options.

Heralded as a "Rising Star of American Cuisine" by the James Beard Foundation, Chef McCabe has cooked for some of Hollywood's hottest stars and epicureans from across the globe. McCabe and his culinary team consistently won awards, including the Golden Sceptre and Golden Bacchus awards from the Southern California Restaurant Writers, the Wine Spectator Magazine Award of Excellence, and Best Hotel Dining by San Diego City Search. In addition, he has been invited on many occasions to cook at the James Beard House in New York.

Before you get started on Chef Paul's amazing recipes, know that they're not so simple; I'm not sure he wants them to be. The time and attention to detail they take will surely be rewarded with your first taste. I gave all chefs featured in *Olive Oil and Vinegar for Life* only two parameters. Each recipe must use an extra virgin olive oil or a Balsamic Fruit Vinegar or both. Only when I arrived at L'Auberge and met Paul did he break the news to me that he was going to create a Corned Beef Tongue. "Uhhh, I'm not sure my readers will enjoy tongue, will you consider making something else?" I asked with (I'm sure) an expression of quizzical discomfort on my face. Chef McCabe looked at me with mischievous eyes as he taunted me, "But you *said* I could make anything I wanted as long as it included olive oil and/or vinegar."

And yes, dear Chef Paul, I did say that. So, dear reader, try it. I believe it will astonish you and your guests with its succulent aroma and flavor.

CORNED BEEF TONGUE, SMOKED OLIVE OIL WHIPPED POTATOES, OLIVE OIL HEN EGG, AND OLIVE OIL CRACKER

SERVES 4

For the Corned Beef Tongue

2 beef tongues (no two tongues weigh the same, so ask for the smallest ones at your local butcher shop)

2 qt cold water

12 oz kosher salt

4 oz brown sugar

¼ oz sodium nitrate

3 bay leaves

2 cloves garlic

2 tsp black peppercorns

2 tsp mustard seeds

1 tsp whole allspice berries

1 tsp dried thyme

1 large onion

2 carrots

2 stalks of celery

Lay tongues in a single layer in a plastic container. Bring water, salt, sugar, and sodium nitrate to a boil in a saucepan.

Remove from heat and cool. Pour brine over tongues, it should cover them by a few inches. Add spices and place a heavy plate on top of tongues to keep them submerged. Cover and place in the refrigerator for 6–8 hours. Remove tongues from brine and rinse under cold water for 2 hours. Remove tongues from water and place them in a large pot. Add onion, carrots, celery, and enough cold water to cover the tongues by a few inches. Simmer tongues for 3–4 hours or until very tender. Remove the skin while still warm. Wrap tongues tightly in plastic wrap and chill overnight. Slice thinly to serve.

Smoked Olive Oil

5 oz Global Gardens Farga Extra Virgin Olive Oil

Put olive oil in a small cup and cover with plastic wrap. Slice a half-inch slit in the top. With a PolyScience smoking gun, place the rubber tube through the hole and into the oil. Place hickory chips in the gun and begin to smoke the oil. Repeat the process about 3 times. Reserve.

Smoked Olive Oil Whipped Potatoes

2 large Yukon Gold potatoes, peeled and chopped

3 c warm milk

5 oz smoked olive oil

salt

Place the potatoes in a small pot with the milk and simmer until tender. Remove the potatoes from the milk and put them into a ricer. Reserve hot milk. Rice the potatoes and whip in one cup of the hot milk and 5 ounces of smoked olive oil. Season with salt to taste and reserve warm.

For the Hen Egg

4 eggs

2 tbsp + 2 tsp Global Gardens Farga Extra Virgin Olive Oil

With a skewer, poke a small hole in the top of the eggs and inject 1 tsp of the olive oil into the egg whites, being careful not to puncture the yolk. Cover the hole with masking tape. Place the eggs in a vacuum bag and seal under medium pressure. Sous-vide the eggs at 62.5° for one hour and reserve.

For the Olive Oil Cracker

(makes lots of crackers that will keep well in Ziploc bags)

3 ½ c all-purpose flour
2 ½ tbsp sugar
1 ½ tbsp salt
4 ½ tbsp Global Gardens Kalamata Extra Virgin Olive Oil
1 ½ cups milk
green onions, chopped as needed

Sift together flour, sugar, and salt. Add olive oil and milk and knead till combined.

Let rest for 1 hour. Roll out the dough in a pasta machine until it's paper thin. Place dough sheets on a sheet pan lined with parchment paper, brush with water, and sprinkle with chopped green onions and smoked coriander mixture. Bake at 375° for 15 minutes or until crisp. Cool and break into free-form shapes.

Presentation

Place the potatoes on a plate and make a small well for the egg to sit in. Remove the eggs from the bag and crack the top to remove the insides. Place on top of the potatoes. Place the sliced tongue on the plate and drizzle with Global Gardens Fig Balsamic Vinegar. Finish the dish with sour grass and flowers.

RAW ORGANIC VEGETABLES, SUN-DRIED TOMATO PUREE, OLIVE OIL CAKE, AND BLOOD ORANGE VINAIGRETTE

SERVES 4

For the Vegetables

1 c each baby carrots, parsnips, beets, radish, and asparagus, shaved thin

For the Sun-Dried Tomato Puree

½ c sun-dried tomato

1 tsp minced garlic

1 tsp salt

1 c tomato juice

1 c olive oil

2 tbsp chopped chervil

Add sun-dried tomato, garlic, salt, and tomato juice to a small sauce pot. Cook until tender. Put tomato mixture in blender and slowly drizzle olive oil into it. Once blended to a smooth consistency, strain through drum sieve. Cool puree then fold in fresh herbs.

For the Dried Olive Oil

¼ c kalamata olives

Preheat oven to 295˚. Dry off the olives with a paper towel. Lay the olives on a half sheet tray. Cook them at 295˚ for 4 hours. Let cool at room temperature. The olives should be brittle enough to crush with your fingers. Put the dry olives into a blender and pulse to a soil texture. Reserve.

For Pumpernickel Crumbs

1 qt loaf pumpernickel bread

Cut the bread into slices. Preheat oven to 295˚. Lay slices of bread on a half sheet pan with space in between each one. Cook bread for 12½ minutes. Take the sheet tray out and flip all sides of bread. Cook for another 12½ minutes. Take out and let cool at room temperature. The bread should be completely dry. Break them into small crumbs and crumble them in a blender. Reserve.

For the Dehydrated Goat Cheese

4 oz goat cheese

Break up goat cheese into small pieces and place on a dehydrator tray (available at most specialty kitchen stores). Let dehydrate overnight. Reserve.

For the Olive Oil Cake

¾ c whole eggs

⅛ c egg yolks

⅓ c milk

3 tbsp olive oil

3 tbsp sugar

pinch of salt

⅛ c flour

Mix all wet ingredients in a blender then add dry ingredients and blend for 1 minute. Transfer to an iSi container (Note from Theo: this is the perfect cream dispenser for the easy preparation of fresh whipped cream and many other cold applications, but the dynamics of the salad are not lost if you don't have access to one), charge with two chargers and set aside. Poke four holes in the sides of a paper cup and disperse the cake in the cup (either by the iSi container or by pouring), not exceeding 3 inches. Set a microwave to 90 percent power and cook the cake for 40 seconds. Turn the cup upside down on a plate and let stand for 1 minute. Tap the bottom of the cup to release the cake. Break into pieces and reserve.

For the Olive Oil Powder

1 c maltodextrin

3 tbsp olive oil

Mix maltodextrin and olive oil in a small bowl until powdery. Reserve.

For the Black Garlic Puree

3 heads black garlic, peeled

2 tbsp hot water

2 tbsp olive oil

¼ c hot water

2 oz olive oil

Peel the garlic and mix the 2 tbsp hot water and 2 tbsp olive oil. Put in a vacuum pack and vacuum seal on high. Let sit for 2 minutes. Take out of the bag and put in a blender. Blend on high and slowly add the hot water and olive oil. Pass through a drum sieve and put puree in a plastic squeeze bottle. Reserve.

Equal parts of the following:
Global Gardens Blood Orange Balsamic Vinegar
Global Gardens California Estate Farga Extra Virgin Olive Oil

Presentation

Spread the sun-dried tomato puree at the bottom of a dinner plate and add the goat cheese, olives, and black garlic. Toss the vegetables in the vinegar and oil and arrange on the plate. Finish the dish with olive oil powder and olive oil cake and serve.

OLIVE OIL GOAT CHEESE MOUSSE, COMPRESSED RHUBARB, OLIVE OIL STREUSEL, DEHYDRATED STRAWBERRIES, AND LEMONGRASS SABAYON

For the Mousse

3 pieces gelatin sheets

1 ½ c cream

⅛ c Global Gardens Kalamata Extra
    Virgin Olive Oil

½ c cream cheese

1–2 ends rosemary

1 c goat cheese

½ c sugar

Bloom gelatin.
Whip cream.
Put olive oil, cream cheese, rosemary, goat cheese, sugar, and bloomed gelatin in a bowl over double boiler.
Once melted and smooth, strain over whipped cream
Fold together.

Pour into prepared pan (½ sheet pan lined with plastic wrap)
Freeze.
Cut in circles.

For the Rhubarb

2 pieces fresh rhubarb, sliced thin

2 tbsp sugar

Toss the sliced rhubarb in sugar and put into a vacuum bag. Seal on high pressure and reserve.

For the Strawberries

Thinly slice a handful of strawberries and dehydrate for 3 hours or until crisp and reserve.

For the Streusel

¾ c sugar

¾ c flour

1½ tsp cinnamon

1 tbsp water

¼ c Global Gardens Kalamata Extra
  Virgin Olive Oil, cold

Mix the dry ingredients and add the water, then fold in the oil. Crumble onto a lined sheet pan and bake at 350˚ for 10 minutes and reserve.

For the Lemongrass Sabayon

1 stalk lemongrass

½ c water

½ c sugar

yolk of 1 egg

½ c whipped cream

Make simple syrup with lemongrass, water, and sugar.
Steep lemongrass and leave out overnight.
Take yolk from egg, add to simple syrup, and whisk over water bath until thick.
Let cool and add whipped cream

Presentation

Remove rhubarb from the bag and place at the bottom of a plate. Add circles of the mousse, strawberries, sabayon, and streusel and serve.

CHEF TIMOTHY RALPHS

Estancia La Jolla Hotel & Spa • 9700 N. Torrey Pines Road •La Jolla, CA 92037 • estancialajolla.com

Chef Timothy Ralphs led the kitchen of L'Auberge's sister property, Estancia La Jolla Hotel & Spa. Estancia is truly a Caliterranean gem, fantastically wrapped in a landscape saturated with mature California tropical and native plants, I thought I was stepping into an older yet pristine California resort. My hostess spoiled me and my guest with massages (à la Caliterranean—outdoors in a fragrant private garden) so that we could immerse ourselves in Chef Ralphs's clean, flavorful spa cuisine, forgetting the traffic-snarled highways nearby and enjoying the sanctuary that is Estancia in totality. Ralphs prepared two superior spa dishes for this book, flawlessly gorgeous with clean, crisp flavors and absolutely pretty to photograph too!

TROPICAL TUNA

5 oz ahi tuna loin

2 tbsp togarashi spice

⅛ each papaya, seeded and sliced

⅛ each mango, julienned

⅛ each avocado, fanned

1 tsp tangerine lace

3 each frisée leaves

1 tbsp Global Gardens Black Currant Vinegar Reduction

¼ each Kaffir lime leaf, julienned

2 each grape tomato, split

1 each passion fruit, split

4 tbsp Global Gardens Black Currant Vinegar

3 each popcorn shoots

Procedure

Trim ahi into a block. Place togarashi spice on a plate and roll the tuna over it to "crust" the tuna.

Over medium heat, sear tuna in 2 tbsp of oil on each side for 30 seconds. Rotate tuna so all four sides are seared—do *not* exceed 30 seconds per side in order to guarantee equal cooking.

Remove tuna from pan and allow to cool before slicing. Slice tuna into 5 1-oz slices.

Plating

Lay out the papaya, mango, and avocado on one side of the plate in a semicircle, allowing each item to overlap on the other. Within the half-moon space, place the tangerine lace that has been lightly dressed with a squeeze of lime juice. Next, toss the frisée with the black currant vinegar and season to taste. Now fold in the Kaffir lime leaf. Place the frisée next to the papaya. To the left of the frisée, lay the shingled pieces of tuna on the plate. Place the halved grape tomatoes in between the tuna and the frisée. Place the half passion fruit in the center of the plate between the papaya and the frisée greens. Remove the passion fruit seeds from the other half and place on the plate next to the sliced tuna. Discard the remaining passion fruit shell. To finish, using a small spoon, drizzle the bottom of the plate with the black currant vinegar reduction. Garnish with popcorn shoots.

OLIVE OIL & VINEGAR FOR LIFE

POTATO AND BEET SALAD

Ingredients (per serving)

- 1 each Yukon Gold potato, peeled
- 2 each baby red beets, roasted, peeled, split
- 2 each baby gold beets, roasted, peeled, split
- 2 each baby candy-striped beets, roasted, peeled, split
- 3 bay leaves, 6 cloves, 9 black peppercorns
- 2 each shiitake mushrooms, roasted, split
- 2 each oyster mushrooms, roasted, split
- 4 tsp herbed goat cheese
- 1 tbsp California Meyer Lemon Balsamic Bliss Vinegar
- petite farmer's greens
- petite amaranth sticks
- 1 tbsp Global Gardens Farga Extra Virgin Olive Oil

Peel the potato and discard the skin. Next, trim the top and bottom of the potato so it is cylinder shaped and will sit flat on a plate. Now, using a small melon baller tool, slowly and carefully remove the inside of the potato, so the potato looks like a cup. Place the potato cup in the bottom of a saucepan. Slowly submerge the potato in cold water. Bring water to a simmer slowly. Simmer potato in water for 12 minutes until soft but not so fragile that it will break once you touch it.

Preheat oven to 375°. Place the baby beets in three separate pans. Add one bay leaf, two cloves, and three black peppercorns per pan. Fill the pans with water until the beets are half under water. Cover the pan with foil and place in the oven. Bake for 35 minutes.

Clean mushrooms and place in a bowl. Toss mushrooms with one tbsp olive oil, salt, and black pepper. Toss until mushrooms are well coated. Place mushrooms on pan and bake in oven for 10 minutes at 375°F.

Plating

Place the warm potato cup in the center of the plate. Arrange the baby beets, cut in half, in alternating order around the potato cup in a circle about 3 inches from the bottom of the potato. Insert the mushrooms into the circle at a consistent distance from each other in between the beets. Place 1 tbsp of herbed goat cheese at a time into the beet circle, with the cheese around the potato cup in between the beets. Next, dress the baby greens with the "Bliss" Vinegar and season to taste. Carefully arrange the baby greens and amaranth sticks, like a floral arrangement, and place in the potato as you would put flowers in a vase. Finally, drizzle the Farga Olive Oil over the beets and mushrooms in the circle.

CHEF BILL WAVRIN

Glen Ivy Hot Springs Spa • 25000 Glen Ivy Road • Corona, California • 888.453.6489 • glenivy.com

Chef Bill Wavrin is a celebrated chef, noted author, food columnist, and television personality. Known for his warmth and genuine, larger-than-life persona, Chef Bill is a regular on the Food Network and a contributor to *Chile Pepper* magazine. Author of a James Beard Award–nominated cookbook, he has earned national recognition for fusing healthy, fresh produce and locally sourced seasonal ingredients into his menus. Chef Bill has been honored for his innovative and pioneering cuisine with a Platinum Carrot Award as one of the "Healthiest Chefs in America" by the Aspen Center for Integral Health.

Chef Bill and his recipes have been featured in over thirty publications, including *The New York Times*, *Town & Country*, *Cooking Light*, *Bon Appetit*, *The New Yorker*, *Vogue*, the *Chicago Tribune*, *Elle*, *Travel + Leisure*, and *Food & Wine* magazine. Chef Bill is the only spa chef to have received a ten-page spread devoted to him in the legendary food magazine, *Gourmet*.

As former executive chef at the noted Rancho La Puerta, Chef Bill earned numerous accolades, including top ratings from *Condé Nast Traveler* and *Travel and Leisure*. He has consulted for Cunard, the cruise line, as well as directed cuisine at the acclaimed Golden Door Spa in California. Long hailed as a triple-crown winner in the culinary community for his work at the three top spas in the world—Miraval, the Golden Door, and Rancho La Puerta—Chef Bill is also a popular TV guest and host, and has been featured on the following:

NBC Today
Good Morning
Good Morning America
Discovery Channel
A&E's "Great Spas of the World"
The Food Network

Not noted in any of Chef Bill's biographical information is how nice he was to me personally, and how much time he devoted to my Global Gardens entourage of spa

groupies when we visited Glen Ivy Hot Springs Spa to taste the fruits of his creative juices and shoot his recipes for this book, while soaking up the delicious, naturally sensuous environment that is California's oldest surviving day spa. I am the first to be reticent to walk around in a swimsuit, robe, and flip-flops all day, but after about thirty minutes of taking in the natural wonderment that is Glen Ivy—the landscape, sounds, sights, and smells—well, it got easy really fast. With nineteen different pool experiences, including mineral baths, lounge pool (changed my life), plunge pool, mud grotto, water aerobics, hot stone massage, the vista spa, and our own VIP cabana where we were served all day by knowledgeable, service-oriented, friendly personnel, the three-day

experience went beyond eating the most tasteful spa cuisine I've ever had. A magical time was had by all, and I look forward to returning.

I asked Chef Bill for just three recipes, he spoiled us with six (and even more, if you count the recipes within the recipes themselves). Thank you, Bill, for your generosities of soulful cooking, the time you took with me in your busy kitchen, and the love that exudes in every morsel of your fantastically serious yet pleasant food. I had a very difficult time typing out these recipes without craving every memorable bite. There seems to be a lot of steps to Chef Bill's creations, and all are absolutely worth every loving second!

APPLEWOOD-SMOKED COPPER RIVER SALMON WITH WASABI AVOCADO AND CILANTRO CREMA

YIELD: 6 SERVINGS

Ingredients

12 baby red potato slices

1 tbsp Global Gardens Ascolano Extra Virgin Olive Oil

1 tbsp Global Gardens Raspberry Balsamic Vinegar

salt and fresh ground black pepper to taste

½ red onion, diced

6 oz wasabi avocado, recipe follows

4 oz cilantro crema, recipe follows

6 oz applewood-smoked salmon

Method

Light grill or preheat oven to 375°.

Step 1: Cut 12 discs from the potatoes. Place in a small bowl and toss the potato discs with the oil and vinegar, salt, and freshly ground black pepper. Allow to marry 5 minutes. Remove potatoes to a papered sheet pan and toast in the hot oven for 20 minutes or until golden brown.

Step 2: Sauté the onions in what's left of the olive oil and vinegar mixture from the potato.

Step 3: Make the crema and wasabi avocado.

Step 4: Presentation on serving plate: Place 2 discs in the center of a salad plate and portion 1 ounce of the salmon onto one tostada and 1 ounce of the wasabi avocado. Garnish with the crema and wedge a sprig of cilantro and a sprinkle of red onion.

Wasabi Avocado

1 avocado, diced

1 tbsp lime juice

2 tbsp tomato, diced

4 tbsp English cucumber, diced

1 dash wasabi oil

1 dash sesame oil

½ tsp pickled ginger, minced

1 tbsp sesame seeds, toasted

Method

Place everything in a bowl and gently mix to combine. Season to taste.

Cilantro Crema

Ingredients

1 c light mayo

1 tbsp lime juice

1 clove fresh garlic, minced

2 tbsp cilantro, finely chopped

Method

Place everything in a blender and blend until smooth.

BLACK BEAN SOUP WITH SMOKED CHICKEN CRÈME FRAÎCHE & MORITA CHILIES

MAKES 8 CUPS

Method

Step 1: Heat the olive oil a medium stockpot over medium-high heat. Add onions, celery, carrot, and garlic. Cook 5 minutes. Add the morita chili, oregano, and cumin. Cook another minute, stirring. Add the vegetable stock and beans. Bring contents to a boil, reduce heat, and simmer for 20 minutes.

Step 2: Carefully ladle half of the soup into a blender. Puree the soup until smooth. Return the pureed soup to the pot and heat to serve. You may need to add a little more stock or water to the soup for a creamy consistency if the soup is too thick.

Step 3: Season to taste. Ladle 6 ounces of the soup in warmed bowls. Place ½ tsp of the cilantro, diced tomato, and diced chicken on top of the soup. Artistically drizzle 1 tbsp of the crema around the soup. Serve straightaway.

Ingredients

1 tbsp Global Gardens Mission/ Manzanilla Blend Extra Virgin Olive Oil

1 onion, diced

½ rib celery, diced

1 carrot, diced

2 cloves fresh garlic, minced

1 whole morita chili

1 tsp fresh oregano

¼ tsp ground cumin

4 c vegetable stock

8 c black beans, cooked

¼ c fresh cilantro, chopped

¼ c tomato, diced

4 oz smoked chicken breast, diced

½ c crema Mexicana

OVEN-ROASTED ACORN SQUASH STUFFED WITH RATATOUILLE OF VEGETABLES AND CHICKEN SAUSAGE

YIELD: 4 SERVINGS

Ingredients

4 acorn squash

1 tbsp Global Gardens Apricot Balsamic Vinegar

1 tsp Global Gardens Ascolano Extra Virgin Olive Oil

1 lb chicken sausage, remove from casings

1 yellow onion, diced

1 zucchini, diced

1 yellow squash, diced

1 eggplant, diced

1 red bell pepper, diced

1 green bell pepper, diced

1 lb crimini mushrooms, diced

2 cloves fresh garlic, minced

2 Roma tomatoes, chopped

1 tsp fresh thyme

1 tsp fresh oregano

1 tbsp fresh basil

½ tsp fresh sage

4 oz Pomodoro sauce

Kosher salt and fresh ground black pepper to taste

Method

Preheat oven 400°

Step 1: Carefully cut around the top of the squash, about ½ inch from the top, making triangular cuts so you have a jagged edge; set the top aside for later use. Slice a thin piece off the bottom of each squash to level them. Clean out the seeds with a spoon. Drizzle the inside of each squash with 1 tbsp of the balsamic vinegar. Oil the outside of the squash. Place the squash with the reserved lids and bake for about 30 minutes or until soft. Set aside until needed.

Step 2: Place a large sauté pan over medium-high heat and add the olive oil to lightly coat the bottom of the pan. Add the sausages and brown lightly for 3–4 minutes. Add the remaining vegetables and cook until the vegetables are al dente, about 2–3 minutes. Add the herbs and the Pomodoro sauce and simmer for 2 minutes. Season to taste.

Step 3: Fill each squash with the sausage mixture and place in the oven for 5 minutes. Remove the squash from the oven and place each one on a warmed dinner plate, placing the lid off center as a garnish.

OLIVE OIL & VINEGAR FOR LIFE

SAUTÉED SCALLOPS IN SWEET PEA SAUCE WITH OVEN-ROASTED GARLIC, TOMATOES, AND BABY BASIL

YIELD: 4 SERVINGS

Ingredients

1 tbsp Global Gardens Koroneiki Extra Virgin Olive Oil

1 stalk thyme

1 garlic clove

4 scallops

salt

pepper

Method

Place olive oil in pan with thyme and slightly smashed garlic clove. When the oil is hot, place the scallops in the pan and allow them to get a nice golden color, turn it, and cook for 10 more seconds then remove scallops from the pan.

Sweet Pea Sauce

YIELD: 2 CUPS

Ingredients

1 tbsp Global Gardens Koroneiki Extra Virgin Olive Oil

1 c chopped yellow onion

1 tsp minced fresh ginger

½ tsp minced fresh garlic

6 c shelled sweet peas

2 c chicken or vegetable stock

2 c fresh spinach, washed thoroughly

Kosher salt and freshly ground pepper to taste

Method

Step 1: Heat a sauce pot over medium heat and add the olive oil. Stir in the onion, ginger, and garlic and cook 2 minutes. Add 2 cups of the peas and cook 1 minute. Stir in the stock and simmer for 5 minutes. Add 3 cups of the remaining peas and spinach and cook 5 minutes. Take off the heat and carefully ladle the hot mix into a blender and process until smooth. Strain the pea mixture through a colander lined with cheesecloth or a fine-mesh strainer into a saucepan. Add the remaining 1 cup of peas and bring to a simmer for 2 minutes. Take off heat and season to taste.

Use the sauce immediately or cool
down in an ice bath. Store in an airtight
container for up to 1 week in the
refrigerator or freeze for up to 2 months.

Oven-Roasted Tomatoes

1 tbsp Global Gardens Koroneiki Extra
Virgin Olive Oil

1 garlic clove, sliced very thin

1 thyme sprig

6 Roma tomatoes, cut lengthwise

salt

pepper

baby basil

Method

Preheat oven to 250°

Step 1: Place olive oil and sliced garlic
with the thyme on a sheet pan. Place the
cut tomatoes upside down, sprinkle with
salt and pepper, and bake at 300° for 40
minutes or until nice and roasted.

Step 2: Place 3 tomatoes in the center of
a soup dish and top with a scallop.
Pour the Sweet Pea Sauce around and
place a pinch of baby basil on top of the
scallop.

83

ROMANA LAMB RACK SCOTTADITO WITH LEMON AND ROSEMARY VINAIGRETTE AND ROASTED MEDLEY OF BABY POTATOES

YIELD: 5 SERVINGS

Ingredients

- 2 racks lamb, fat trimmed off and Frenched
- 2 tbsp Herb Garlic Rub, recipe follows
- 4 small baby red potatoes, quartered
- 4 fingerling potatoes
- 1 small onion, julienned
- 8 cloves garlic
- 2 tbsp Global Gardens Mission/ Manzanilla Blend Extra Virgin Olive Oil
- 1 tbsp rosemary, chopped
- 1 tsp thyme, chopped
- 2 lemons, cut into wedges
- 4 tbsp Global Gardens Mission/ Manzanilla Blend Extra Virgin Olive Oil
- 1 c lemon juice
- 4 cloves roasted garlic, minced
- 2 roasted shallots
- 2 anchovy fillets
- 1 tsp fresh rosemary, minced
- 1 tbsp lemon zest

Method

Preheat oven to 425°.

Step 1: Clean racks of fat and French bones. Season cleaned racks with Herb Garlic Rub. Wrap in plastic wrap and set aside in the refrigerator until needed. Will last 5 days in the fridge.

Step 2: Toss the potatoes with the onion, cloves of garlic, rosemary, thyme, and lemon wedges and roast in the oven 25–30 minutes or until cooked through.

Step 3: Place the olive oil, lemon juice, roasted garlic, shallots, anchovies, and rosemary in a blender and blend until smooth. Add the zest and place in a squeeze bottle and refrigerate until needed.

Step 4: Season the racks and sear in hot sauté pan in the oil over high heat until all sides are well browned. Place in a hot oven for 8 minutes. Remove from oven and place on a plate 2–3 minutes so racks can relax.

Step 5: Place 2 potatoes, quartered, in the center of a warmed dinner plate. Slice the rack into three chops (3 bones) and stand 2 chops (2 bones) with one bone interlocking. Spoon 2 ounces of the sauce over the lamb.

Fresh Herb Garlic Rub

YIELD: ½ CUP

Ingredients

- 1 tbsp fresh basil, chopped
- 1 tbsp fresh oregano, chopped
- 1 tbsp fresh tarragon, chopped
- 1 tbsp fresh thyme, chopped
- 1 tbsp fresh parsley, chopped
- 1 tbsp fresh sage, chopped
- 8 cloves garlic, minced
- ¼ c Global Gardens Mission/Manzanilla Blend Extra Virgin Olive Oil

Method

Step 1: Wash, dry, and chop the herbs. Place in a blender with the garlic and oil and pulse to combine. Place in a sealable jar and set aside in the refrigerator.

STRAWBERRIES TOSSED IN RASPBERRY BALSAMIC SIMPLE SYRUP WITH SERRANO ESSENCE

This is a great party appetizer. The combination of the berries, brown sugar, and chili are a perfect eye-opener. The addition of the Global Gardens Raspberry balsamic adds that subtle touch that will leave you wondering. Yum!

YIELD: 4 SERVINGS

Ingredients

4 large strawberries, washed

12 blackberries, washed

12 raspberries

Serrano Simple Syrup, recipe follows

2 tbsp Global Gardens Raspberry Balsamic Vinegar

6 mint leaves, finely chopped

Method

Step 1: Wash and stem the strawberries and blackberries. Cut ⅛ inch from the bottom of each berry so it may stand alone without tipping over during service. Set aside in the refrigerator until needed.

Step 2: Place the strawberries in the center of each plate. Surround the strawberries with 3 each of the raspberries and blackberries. Drizzle each plate with the Serrano Simple Syrup. Sprinkle a few drops of the balsamic vinegar over the strawberries. Garnish with a sprinkling of mint. Serve and enjoy!

Serrano Simple Syrup

YIELD: 1 CUP

Ingredients

4 serrano chilies

1 qt of water

½ c sugar

2 tbsp white balsamic vinegar

Method

Step 1: Wash and stem the chilies and slice into thin rings. Place a sauce pot over medium-high heat with the water and bring to a boil. Blanch the chilies in the water for 30 seconds. Remove to a plate lined with a paper towel. Pat the chili rings dry and remove most seeds.

Step 2: Place a small saucepan over medium heat with the sugar and vinegar. Cook until the sugar starts to boil. Reduce to a very slow simmer and cook 5 minutes. Remove the syrup from the heat and stir in the chilies. Pour into a sealable container and allow to cool. Set aside until needed.

Caliterranean

BREAKFAST

Dawn—my favorite part of the day. I always sleep with my curtains open so I can enjoy the first hint of light painting my backyard with pastel tones. My own Caliterranean lifestyle morning routine begins with a series of Tibetan chakra balancing exercises that take about twelve minutes. A nice protein-inspired breakfast will carry me through midday. The special breakfast recipes included here are tried-and-true originals. I look forward to receiving your suggestions too.

STUFFED FRENCH TOAST

Now *here* is an opportunity to get creative! Use your favorite greens for wilting and veggies for sautéing—the olives and feta cheese will balance the flavors of anything fresh. You can even add seedless grapes, sliced in half, or seedless tangerine sections from pixies or clementines. This makes a perfect brunch dish—my favorite treat to have on a late Sunday morning after a brisk walk with the dog, enjoying the early light.

PER SERVING

Ingredients

- 2-inch slices fresh French bread (not to be confused with a baguette)

- 1 egg, beaten and poured into a 6- to 8-inch dish

- 1 tbsp Santa Ynez Italian Varietal Blend Extra Virgin Olive Oil

- 2 tbsp each minced red peppers, purple shallots, seedless kalamata olives, and, feta cheese crumbles (I like FAGE feta best)

- 1 c spinach

Method

Soak French bread evenly with entire egg on both sides.

Cook

Add olive oil to skillet over medium heat. Sauté red peppers and shallots in olive oil until shallots are a little brown, add spinach and stir until wilted. Remove from heat and place into a warmed bowl. Now brown bread on both sides in the same skillet that has been seasoned by the olive oil and veggies, about 2 minutes on each side. Remove bread, slice in the middle, making a pocket (don't slice all the way through), and place onto serving dish. Add veggies to warm skillet and toss in olives and feta cheese just until feta starts to soften, about 1 minute. Stuff veggie mixture into French toast pocket and enjoy immediately.

Note

You can make plain French toast using this egg and olive oil combination without stuffing; keep in freezer bags for a quick toaster-thawed breakfast. Kids love it!

LEMONETTE & BERRY SEXY PANCAKES

This recipe is adapted from a Global Gardens monthly recipe winner, Nik Weinberg-Lynn. What a fantastic, healthy alternative to regular, boring pancakes! I have made these with gluten-free flour—Bette's Four Flour Blend—which adds fiber and protein. Shown here with Caliterranean Syrup.

SERVES 3–4

Ingredients

1 organic egg

1 c plain yogurt (I like whole FAGE Greek Yogurt)

2 tbsp Meyer Lemon Extra Virgin Olive Oil

2 tsp fresh lemon juice (juice after zesting)

½ c organic all-purpose flour (I have substituted Bette's Four Flour Blend successfully)

½ c quick-cooking oats

2 tbsp organic sugar

1 tsp baking powder

½ tsp baking soda

¼ c + 2 tbsp milk

⅛ tsp salt

½ c blueberries (or your favorite berries)

zest of 1 lemon (Meyer, if you can get them)

Method

Beat egg, yogurt, olive oil, and lemon juice together well. Add remaining ingredients and stir together.

Cook

Pour about ¼ cup of batter on lightly oiled, hot griddle. Turn when pancakes are covered with bubbles. Makes about 12 3-inch pancakes.

CALITERRANEAN SYRUP

What a great surprise this protein-healthy syrup is—it adds a rich, flavorful dimension to French toast, pancakes, and desserts.

Tip: Don't cook this ahead of time—create just prior to serving. If it thickens too much, heat again and serve. It doesn't work at room temperature like regular pancake syrups.

SERVES 4

Ingredients

- ½ c organic sugar
- ¼ c Ascolano or Koroneiki Extra Virgin Olive Oil
- 1 organic egg, beaten

Method

Combine all ingredients in a small skillet into a paste.

Cook

Stir ingredients together constantly over low-medium heat, about 10 minutes.

SAY CHEESE! PUMPKIN STREUSEL DOUGHNUTS

I prefer breakfast breads and pastries when you can taste the ingredients, not the sugar. Everything pumpkin pleases me, especially when paired with just the right cheese, creating a comforting texture. I'm not gluten intolerant, but a lot of my friends are, and these doughnuts are lighter and more intent on the nuanced flavors of the pumpkin, cheese, and spice blend because of the flour used.

YIELD: 24 DOUGHNUTS

Ingredients for Doughnuts

2 ¼ c Bette's Gourmet Four Flour Blend

2 c organic sugar

½ tsp salt

1 tbsp baking powder

1 tsp cinnamon

1 tsp ground nutmeg

2 eggs

1 ¼ c organic canned pumpkin *or* one 6-inch Cinderella baking pumpkin

⅓ c Global Gardens Ascolano Extra Virgin Olive Oil

2 tsp vanilla

Method

If starting out with a fresh pumpkin (delicious!), cut in half and remove all seeds. Rub generously with olive oil and add a dusting of brown sugar. Bake in 400° oven for 30–35 minutes until soft when pierced with fork. Let cool. Scoop out and measure flesh.

(Bonus Recipe: Spread any extra pumpkin onto a shallow baking dish, covering with fresh-sliced Mission figs and 4 oz chèvre or mild feta cheese. Sprinkle with minced fresh rosemary, salt, and pepper, low-broil just until cheese melts.)

Mix dry ingredients together with a fork in a large bowl. Add eggs, pumpkin, olive oil, and vanilla. Set aside.

Ingredients for Cheese Filling

8 oz chèvre or mild feta cheese

1 egg

1 tsp vanilla

3 tbsp sugar

Method

Blend all ingredients together in a medium bowl and set aside.

Ingredients for Streusel

- 5 tbsp Bette's Gourmet Four Flour Blend
- 5 tbsp organic brown sugar
- ½ tsp cinnamon
- ½ tsp five-spice powder
- 3 tbsp Global Gardens Ascolano Extra Virgin Olive Oil
- ¼ c finely chopped organic raw almonds

Method

Blend all ingredients in a small bowl.

Fill lightly oiled doughnut pans about halfway with doughnut batter. Put cheese mixture into Ziploc bag (or pastry bag with a medium point if you have one). Cut a ⅛-inch diagonal from one corner of your bag and squeeze about 1 tbsp of mixture, following a circular pattern, on top of doughnut batter. Sprinkle each doughnut with about 1½ tbsp streusel mix.

Bake at 375° for 20–25 minutes.

CREDIT: RON BOLANDER

FIG 'N' EGG OMELET

The great folks at the Farmhouse Inn created my recipe for me to shoot within their lovely edible flower garden, sharing their own recipes for the lightest scones and muffins—both featuring extra virgin olive oil instead of butter. See? All your life you thought you would need butter for rich, lip-smacking baked goods. Now, lighter and even tastier baked goods are possible—for life—when baked the Caliterranean way!

SERVES 1

Ingredients

2 farm fresh organic eggs
1 tbsp Global Gardens Fig
    Balsamic Vinegar
dash of salt
1 tsp Global Gardens Meyer
    Lemon Extra Virgin Olive Oil
1 oz Havarti cheese
1 tbsp fresh chives

Method

Beat eggs with vinegar and salt. Pour into small skillet (I like my smallest Calphalon for making a perfect omelet every time) coated with lemon olive oil.

Cook over medium heat using a silicone spatula to lift edges of egg, allowing any liquid to spill into the sides, cooking the bottom of the egg until lightly brown. Flip egg over in pan, then add cheese and chives to the middle of the circular area. Overlap egg and remove from heat.

Present with fresh chives and edible flowers. The contrasting flavors of the Havarti and the spiciness of the edible flowers are a great way to say good morning to your palate.

The Farmhouse Inn's Citrus Olive Oil Muffins make an artful, tasteful pairing to the omelet.

CITRUS OLIVE OIL MUFFINS

Bonus! The friendly staff in the Farmhouse Inn kitchen liked the idea of reductions on baked goods, so they shared a muffin recipe I did not expect. The results are a light crumb, not uncommon with olive oil baking. The real surprise came in its ability to balance so many citrus flavors with the nuttiness of the almonds and other ingredients; quite pleasant and perfect with my double macchiato.

YIELD: 12 MUFFINS

Ingredients for Muffins

1 ¾ c all-purpose flour
2 tsp baking powder
½ tsp salt
1 c sugar
4 large eggs
2 tsp orange zest
2 tsp lemon zest
2 tbsp Global Gardens Strawberry
   Balsamic Vinegar
2 tbsp milk
¾ c Global Gardens Ascolano Extra
   Virgin Olive Oil
⅔ c sliced almonds, toasted and roughly
   chopped

Ingredients for Reduction Glaze

2 c strawberry balsamic vinegar

Method for Muffins

Sift together the flour, baking powder, and salt. In a stand mixer, whisk together the sugar, eggs, and zests until light and fluffy, about 3 minutes. Add in the vinegar and milk. Gradually add in the oil. Next, add in the sifted dry ingredients and mix until incorporated. Fold in the crushed almonds. Divide the batter evenly into lined muffin cups.

Bake at 350˚ for 20 to 25 minutes. Allow muffins to cool, then top with vinegar reduction.

Method for Reduction

Reduce vinegar to 50 percent by boiling over medium heat for about 5 minutes in a skillet. See reduction recipe, page 27.

FARMHOUSE OLIVE OIL SCONES

And they said it couldn't be done. The head chef at the Farmhouse shook her head and said she just didn't "know how good they'd be." Everyone in the kitchen looked at me like an intruder, wanting to get on with their busy breakfasting. Check it out! I'm a lover of scones, and I've never tasted one better; did someone say better than butter? For sure, baking with extra virgin olive oil is better for you and better in flavor.

YIELD: 12 SCONES

Ingredients

2 ¾ c all-purpose flour
1 tbsp baking powder
½ tsp baking soda
½ tsp salt
½ c currants
1 c buttermilk
½ c Global Gardens Santa Ynez Italian Varietal Blend Extra Virgin Olive Oil
½ c sugar
1 large egg

Method

Combine flour, baking powder, baking soda, and salt. Sift mixture three times. Stir in currants. In a separate bowl, whisk together the buttermilk, olive oil, sugar, and egg. Add wet ingredients to dry ingredients. Mix just until a soft dough forms. Divide dough in half and form each half into a 6-inch disk. Wrap each disk in plastic and refrigerate for 30 minutes. Remove plastic wrap after refrigeration time. Using a sharp floured knife, cut each disk into 6 wedges. Place wedges 3 inches apart onto a parchment-lined baking sheet.

Bake at 400˚ for, 12 to 15 minutes, or until they are firm but not dry. Serve warm.

OLIVE OIL & VINEGAR FOR LIFE

Caliterranean

BEGINNINGS

The following collection of recipes *can* be created as beginnings, endings, or in-betweens—but as a group, they are *perfect* as appetizers, small plates, or snacks. Besides the cocktail, all of these dishes can easily be made into mains, as I will suggest where fitting (Popcorn for dinner? You'll see!) Whether you try these as beginnings or endings, I am hopeful you will find your own way to enjoy these recipes by experimenting beyond my suggestions.

CALITINI

What festive, tasty fun it was to come up with a new cocktail using Balsamic Fruit Vinegars! My quest took me to many California towns where I met other food purveyors as well as the guys who made this awesome cocktail tray to display my new Calitini. ASI (ArtStyleInnovation.com) is a fun gallery featuring contemporary, one-of-a-kind, custom-designed acrylic furnishings and home accessories. This groovy serving tray comes with matching (or not) neon acrylic coasters and floating bowls for pool flowers, salads, or snacks. See Kicky Veggie Melee and Trail Muscle Mix on pages 126 and 200 for more colors and designs.

1 large or 2 small ice cubes made from any Global Gardens Balsamic Fruit Vinegar. Shown are Mango and Strawberry. Apple Ginger works great too.

regular ice cubes

fresh sprig mint

1 oz Ketel One Vodka (my favorite for mixed drinks)

6 oz tonic water

Vinegar—shown are Mango and Strawberry.

Double the recipe for a tall glass. Place vinegar ice cubes into glass first, then fill glass ¾ full with regular ice. Crush mint leaves to release their natural oils and place the sprig on the ice. Cover with vodka and tonic. Put your sunglasses on, go outside, relax, and rejoice in the outdoors, Caliterranean style!

PRIMO POPCORN

C'mon, popcorn in a cookbook? I'm sorry, but nobody makes better popcorn than me, with every single kernel popped in the bottom of the bowl. Believe it or not, there is a precise methodology for making the best popcorn. Of course, you don't have to use the Meyer Lemon Extra Virgin Olive Oil, but it sure tastes primo!

Ingredients

1 tbsp Global Gardens Meyer Lemon Extra Virgin Olive Oil (use 2 tbsp if you want a more self-indulgent version—after all, you're skipping the butter!)

¼ c organic popcorn

2 tsp Caliterranean Garden Blend

Method

I love my Calphalon pan with the glass lid for this recipe. First, put the olive oil in the bottom and turn heat to medium-high. Wait 30–45 seconds, then add the popcorn, swirling the pan around until each kernel has some oil on it. Place lid onto pan and wait until the corn starts to pop, filling the entire bottom of the pan (or if you don't have a glass lid, wait for about 15 seconds after the corn starts popping). Slide—*don't shake*—the pan, back and forth along the burner, just a few times. Wait until the pan is about half full before doing this again, or about another 20–30 seconds. As soon as you hear the popping slow way down to just a few pops here and there, turn off the heat but don't remove the pan until the last kernel is popped. Sprinkle Garden Blend, immediately upon opening lid, onto hot popcorn. If you're adding another tablespoon of olive oil (super primo!), do that prior to seasoning.

AMOROUS AVOCADO SOUP

I served this soup in small cups as an appetizer on Thanksgiving Day. The spiciness from the roasted jalapeño combined with the creamy, comforting texture is just enough to get your palate started for the big day. This recipe stores beautifully (without turning brown at all) in the refrigerator in your glass blender container for two full days. We've used leftovers as tasty taco fillers with roasted corn, tomatoes, onion, and cilantro. Add chicken or shrimp, or just keep it all veggies. The corn is a wonderful flavor enhancer with the avocado and can also be used to garnish the soup if you want to make it a more filling main dish.

Ingredients

- 1 3- to 4-inch jalapeño
- 4 medium Haas avocadoes
- 2½ c organic chicken broth
- ¼ c Global Gardens Meyer Lemon Bliss Vinegar
- 1 tsp ground cumin
- 1 tsp vanilla extract

Method

Roast jalapeño on all sides in a 400° oven or skillet over medium-high heat until the skin bursts. Remove from heat, peel, and deseed with a fork in each hand so you won't get stinging essences from the pepper oils onto your skin. Peel the avocadoes and place two in a blender with all remaining ingredients. Blend until smooth and add remaining avocadoes. Serve at room temperature or chilled. Either way, you'll fall in love with this Caliterranean enigma.

109

Colossal Garlic

CALIFORNIA ORGANIC

GLOBAL
GARDENS

Perfect For Roasting

ROASTED GARLIC SOUP

I love this soup. Daniel made it for me on a wintry November evening in Santa Barbara—we had a fire in the fireplace, accompanied by an olive oil–grilled baguette and a bottle of Buttonwood Sauvignon Blanc. Whenever I'm feeling chilly or in need of comfort food, that evening . . . and this soup . . . always comes to mind.

Ingredients

2 qt organic chicken broth

1 c Buttonwood Sauvignon Blanc

4 giant cloves Global Gardens Organic Elephant Garlic

¼ c Global Gardens 100% Mission Extra Virgin Olive Oil

¼ tsp salt

1 tbsp finely chopped lemon thyme leaves

2 tbsp chopped chives

Method

This soup is so easy, it just tastes difficult! Bring chicken broth and wine to a boil in an uncovered heavy pot (like Le Creuset) over medium-high heat. Reduce for about 2 hours into almost half the liquid. Meanwhile, brown the garlic over medium heat in a Le Creuset or other heavy skillet with the olive oil and salt, until the garlic is very soft. Set aside until broth and wine have reduced and cooled to room temperature. Using a Cuisinart, blend ½ liquid and garlic mixture. (If you do this when the soup is hot, be very careful, as hot liquid expands quickly in the Cuisinart—only blend 1 cup at a time under this circumstance.) Finish blending and strain any remaining lumps of garlic cloves from the liquid. Return to heat when ready to serve (keeps well refrigerated for 2–3 days), stirring in thyme leaves. Garnish with fresh chives and chive blossoms *or* edible flowers.

BIG SUR TUNA

I am not one for bumper stickers, but my car does sport a small pink oval sticker in the shape of a surfboard that spells the name of my favorite quick getaway in swirling oceanlike letters—Big Sur owns a special place in my heart. Working up an appetite is easy here, with phenomenal hiking, redwoods, the Big Sur River, dramatic surf, and colorful mountainsides that reach directly into the Pacific. This tuna travels well in a cooler, made with olive oil and vinegar instead of mayo. Pears have the highest fiber of any fruit, adding dimension and interest to this healthy, appealing recipe.

Ingredients

- ¼ c minced, drained seedless kalamata olives
- 6 oz can American Tuna (pole-caught in the USA! See americantuna.com)
- ¼ c diced, firm Bartlett pear
- 2 tbsp scallion greens
- 2 tbsp diced tomatoes (organic cherry tomatoes work great off season)
- 1 tsp fresh rosemary
- 2 tsp orange rind
- 2 tbsp Global Gardens Blood Orange Extra Virgin Olive Oil
- 1 tbsp Global Gardens Blood Orange Dark Balsamic

Method

Blend ingredients in a large bowl and scoop onto fresh rolls or salad plates. Drizzle a little extra olive oil and vinegar on the bread if it is very crusty for added flavor and metabolizing factor. Easy Protein Crackers (see page 119) are a delicious accompaniment if serving as a salad.

COWGIRL BLUE CAVIAR
(COWBOYS AND COWKIDS LOVE IT TOO!)

SERVES 4

Picnics become even better when you're thinking Caliterranean. Lovely day, light streaming through trees and sparkling onto wildflowers, great friends, and tasty picnic food all make for a superior afternoon. The avocado will turn a bit bluish when you add the beans, but the symbiosis of flavors will transcend your blues to jazzy pleasure on your palate.

Ingredients

½ cup roasted corn (2 ears)

1 tbsp Global Gardens 100% Mission Extra Virgin Olive Oil

⅛ tsp fine sea salt

3 tbsp Global Gardens Meyer Lemon Bliss Vinegar

1 tsp crushed Global Gardens Elephant Garlic

⅛ tsp freshly ground pepper

1 tbsp chili powder

⅓ c packed minced cilantro

½ c black beans

1 c diced firm Haas avocado

Method

Oven roast the corn on the cob by rubbing first with olive oil and salt, then placing on a cookie sheet in the oven at 400˚, turning after 10 minutes, roasting for 20 minutes total. Cut corn away from cob and put ½ c into a medium bowl. Add vinegar, garlic, spices, and cilantro, then stir together. Add beans, then avocado, and toss lightly so as not to crush the avocado and beans. Serve with tortilla chips.

OLIVE OIL & VINEGAR FOR LIFE

AWESOME APPETIZER SPREAD

It's all about balancing protein with tastes and textures that will fulfill hunger pangs when you're craving a terrific snack. Aficionados of raw cooking can make this easily transportable appetizer spread without roasting the pepper, although I call for it to be done this way because it enhances the flavor of the remaining ingredients.

Ingredients

1 large sweet red pepper

2 tbsp + 1 tsp Global Gardens Mission/ Manzanilla Blend Extra Virgin Olive Oil

½ c minced, drained, pitted kalamata olives

½ c minced, drained, pitted sevillano olives

½ c chopped fresh basil

⅔ c raw organic almonds

¼ tsp salt

1 tbsp Global Gardens Fig Balsamic Vinegar

1 tsp lemon thyme

1 tbsp lemon zest

1 tsp minced Global Gardens elephant garlic

Method

Cut pepper in half, remove seeds, and rub with 1 tsp olive oil and a sprinkling of salt. Oven-roast both sides for 10 minutes or heat in skillet until the skin bursts. Cool and peel pepper, then chop into medium bowl, stirring in remaining ingredients. Can be made ahead of time and saved 2–3 days at room temperature. (There is nothing that will spoil, and refrigerating will take away the texture of the almonds and coagulate the olive oil.) Enjoy as a topping for Easy Protein Crackers (see page 119), pasta, baked potatoes, and salad, as well as for stuffing seafood, chicken, or butterflied leg of lamb.

EASY PROTEIN CRACKERS

I enjoy snacking a lot, but I don't like the empty carbohydrates typically associated with snacks. This cracker resulted from several attempts to make a snack that could taste good all by itself, or even better with something like my Awesome Appetizer Spread (see page 118). The texture is much like that of those English wheat tea biscuits I enjoy so much; these just happen to be a lot better *for* you. You can also make these gluten free with Bette's Gourmet Four Flour Blend. Start with ¼ cup water and add 1 tbsp at a time if baking this recipe gluten free. Humidity and the weather may require less water when using Bette's flour.

Ingredients

1½ c organic Red Mill wheat flour

1 c finely ground organic walnuts

1 tsp fine-grain sea salt or Caliterranean Garden Blend

½ c warm water

2 tsp organic vanilla extract

⅓ c Global Gardens Ascolano Extra Virgin Olive Oil

Method

Mix flour, nuts, and salt together well by hand. Using an electric mixer, slowly add liquids. Dough will be very thick. Form 3 balls with the dough. If it's too dry, pat with water; if it's too sticky, pat with flour. Roll the dough out with a floured rolling pin to ⅛ inch thickness and cut into shapes with a cookie or biscuit cutter. Place onto an ungreased cookie sheet.

Bake at 450° for 7–8 minutes, until bottom is brown.

KALAMATA FETA PISTACHIO ROLLS

This recipe was conceived years ago when I was in a pinch for some appetizers to take to a party and didn't want to go to the grocery store. Being the good Greek girl I am, I usually have a pound of phyllo in the freezer. I *always* have olive oil, and the rest came together with what I found in the pantry. These freeze well after baking for 15 minutes. Finish last 10 minutes of baking in a 400° oven after removing from freezer. Place directly onto cookie sheets and bake immediately—frozen never tasted so fresh!

MAKES 80 PIECES

Ingredients

- 1 lb phyllo

- 1 c Global Gardens Greek Kalamata or Koroneiki Extra Virgin Olive Oil

- 1 9-oz jar Global Gardens Santa Barbara Appetizer Spread (*or* 1 c minced, drained kalamata olives, blended with 2 tsp dried Greek Oregano and 1 tsp minced garlic)

- 2 c crushed pistachio nuts

- 1 c feta cheese crumbles

- 80 tiny basil leaves for garnish

Brush olive oil onto five sheets of phyllo, evenly stacking one on top of the other, using parchment paper under the first layer. Spread about 2 tbsp appetizer spread or olive mixture over this layer. Sprinkle with ¼ c pistachios and ⅛ c feta crumbles. Roll the entire layered rectangle tightly. Seal edge with olive oil. Slice each roll with a very sharp, unserrated knife to about 1 inch thick. I get 8 good pieces per roll, discarding the unsightly ends or baking them and setting aside for family snacking. Repeat this process seven more times.

Place onto ungreased cookie sheet and bake approximately 25 minutes at 350° until just slightly brown. Make the rolls the day before your party and store in a foil-wrapped pan. Storing in plastic will make the phyllo soggy. Flavors will meld together more interestingly when made the day before and served at room temperature, but there is nothing wrong with fresh, hot phyllo appetizers if you get your timing right!

Mango Reduction Glaze for Garnish:

Ingredients

1 c Global Gardens Mango Champagne Vinegar

Method

Reduce vinegar by bringing it to a boil over medium heat until product thickens naturally from the sugars in the fruit. Do not overcook; remove from heat as soon as the product reduces to 50 percent because it will get thicker as it cools.

Optional sauce ingredients (use in place of the Mango Reduction Glaze)

2 tbsp San Marcos orange blossom honey

½ c fresh squeezed orange juice

½ c Buttonwood Sauvignon Blanc

orange zest

Method

Blend all ingredients with an immersion blender.

Presentation

Put ½ tsp glaze onto the top of each sliced roll and garnish with basil leaf placed in the middle of the glaze. The basil is critical to the flavor profile of this delectable appetizer.

CALI CRUNCHIES

I grew up eating Fritos thrown on to piping hot Campbell's Tomato Soup with little chunks of cheddar cheese thrown in for flavor. Fritos, in my opinion, offered the perfect crunch to satisfy my cravings for salt and texture. Sorry, Fritos, we're through; move over and make way for my version of the most tastefully edible, healthy crunch around. Kale has a whopping dose of vitamins K (1,327 percent of the recommended daily allowance in one cup!), A (192 percent), and C (over 88 percent), supporting our bones, nervous system, and every body organ in one way or another. It's easy to grow in almost every climate and does well in containers too.

Ingredients

- 1 c fresh kale (large spinach and chard leaves work too)
- 2 tbsp Global Gardens Meyer Lemon Extra Virgin Olive Oil
- 2 tsp Mendocino Sea Salt Crystals

Method

After washing leaves thoroughly, make sure they are completely dry and free from any moisture or water or else this recipe will not work. Diligently spread a very thin layer of olive oil onto both sides of each veggie leaf with a wide pastry brush. Make sure to coat the edges and not to allow oil to pool inside any crevices. Sprinkle with sea salt. Place leaves (touching is OK but not overlapping) on a large glass baking dish or cookie sheet.

Bake at 225° for 16–18 minutes until crispy.

STUFFED VEGGIE FLOWERS

Nasturtiums, zucchini, and patty pan squash blossoms are each ideal for this recipe. Color, flavor, texture, and drama on the dish and palate exude from bountiful gardens all summer long in almost every state; nasturtiums grow wild in California! These stuffed gems make incredible additions to any pasta dish as a main feature with some seafood tossed in (or *not*) and the perfect side dish as shown with the Smoked Mendo Tri-Tip and Cali Crunchies shown all together on the same plate, page 171.

Ingredients

- 6 each nasturtium, zucchini, and patty pan squash blossoms
- 1 c finely grated organic cheddar (I use Clover Farms mild white)
- 1 tbsp chopped fresh thyme
- 2 tbsp fresh chives
- 3 tbsp Global Gardens Meyer Lemon, Ascolano, or Koroneiki Extra Virgin Olive Oil
- ½ tsp Mendocino Sea Salt Crystals

Method

Harvest flowers in the morning or when the sun is at its highest, when blooms are wide open. Don't wash the flowers or they will absorb too much water and will not be good for this recipe. Pull stamens out from squash blossoms, being careful not to get stung from any honeybees hiding in their midst (happens to me at least once a year). Mix finely grated cheese with herbs and stuff by the spoonful into flower heads. Fold the flower petals, forming a closure onto each blossom. Coat lightly with a wide brush imbided with olive oil. Sprinkle with sea salt. Place closely but not touching on a cookie sheet or glass baking dish.

Roast in a 350° oven for 5–6 minutes until cheese just begins to melt and caramelize onto pan. Remove with metal spatula when just starting to cool and serve immediately.

KICKY VEGGIE MELEE

Lotusland in Santa Barbara is a favorite destination when I want a quiet, aesthetic afternoon inspired by light on the most unusual yet natural plant formations, flowers, and water features. This dynamic French chateau property was purchased in 1941, developed over forty-three years by the eclectic Madame Ganna Walska, a well-known Polish opera singer and socialite, making the perfect backdrop for the next three salad recipes featured in this cookbook. Outstanding educational programs at Lotusland serve our community with sustainable botanical practices, imported succulents, and unusual rare plants from around the world, juxtaposed to a most original setting of garden art and architectural artifacts. A favorite for photographers, painters, and lovers of nature, and a must-see even for the casual observer. Taking a stroll around this masterful land transports the senses to Caliterranean heights of glory.

Ingredients for the Salad

1½ c roasted organic corn (3 ears)

1 tbsp Global Gardens California Koroneiki Extra Virgin Olive Oil

⅛ tsp salt

1 head romaine lettuce, chopped

1 cucumber, sliced into ⅛-inch pieces

1 15-oz can garbanzo, kidney, or pinto beans

12–15 cherry tomatoes, halved

½ c packed cilantro

½ medium onion, minced

1 firm Haas avocado, chopped

3 hard-boiled eggs, quartered

Method

Oven-roast corn by rubbing with olive oil and salt, placing in a 400° oven for 20 minutes, turning after 10 minutes. Cut corn from cob when finished. Mix all ingredients, except the avocado and eggs. Lightly toss in the avocado so it doesn't get mushy, and then garnish the top with quartered eggs.

Ingredients for Salad Dressing

- ½ c Global Gardens California Koroneiki Extra Virgin Olive Oil
- ¼ c Global Gardens Meyer Lemon Bliss Vinegar
- 1 tbsp + 1 tsp chili powder
- 1 tsp cumin
- ½ tsp fine sea salt
- ⅛ tsp fresh ground pepper
- 1 clove Global Gardens Elephant Garlic minced
- ¼ tsp sweet paprika or smoked Spanish pimento

Method

Thoroughly mix all ingredients with an immersion blender and pour over salad. This salad makes a hearty main dish with plenty of protein, flavorful sensations, and balanced nourishment everyone will enjoy.

STRAWBERRY FIELD SALAD

Fruit is a delicious enhancement to green salads and can easily turn boring green veggies into a dynamic main dish when accompanied by cheese or nuts for protein. I am showing this salad as an appetizer portion, but we have had it on many nights as a main dish served with Easy Protein Crackers for depth and compatibility.

SERVES 4

Ingredients for the Salad

2 c organic mixed greens or arugula (whatever you do, make sure your greens at least include some arugula, which is a must-pair spicy ingredient with the sweet strawberries)

1 c sliced strawberries

20 ⅛-inch slices cucumber

½ thinly sliced purple onion

4 oz thinly sliced hard cheese like pecorino or kasseri

fresh ground pepper

Method

There is an art to laying out a salad, but it's easy to start with reference to the adjacent photo. Caliterranean cooking shares the art of color, texture, and design on the plate; have fun with it, and you will be surprised at the comments and flavorful results. After arranging veggies, top with cheese and fresh ground pepper to taste

Ingredients for Dressing

¼ c Global Gardens California Koroneiki Extra Virgin Salad Dressing

2 tbsp Global Gardens Strawberry Balsamic Vinegar

⅛ tsp fine sea salt

Method

Blend ingredients thoroughly with an immersion blender until creamy and pink. Drizzle over salad just before serving.

AVOCADO OLIVE SALAD

Sometimes there are dishes that are so easy to prepare, I'm embarrassed to include them in a cookbook. This salad is a great example of Caliterranean display, flavor, and simplicity. A favorite for friends and family, taking only minutes to prepare.

Ingredients

2 firm Haas avocadoes, seeded and sliced into ⅛-inch pieces, keeping the entire width of the avocado intact

1 large purple shallot, thinly sliced

16 Global Gardens Thassos Oil Cured Olives, pitted

½ large organic red pepper

lemon slices for garnish

4 tbsp Global Gardens Meyer Lemon Extra Virgin Olive Oil

fresh ground black pepper

Method

Creatively place salad items on colorful dish. Drizzle with olive oil and grated fresh pepper.

MUSHROOM PARADISE PÂTÉ

Well, if this doesn't impress friends, family, and chefs, nothing will. I am a huge fan of charcuterie and "real" chopped liver or duck liver pâté. Call me a heathen. In celebration of wild mushroom season, this recipe truly tops the goose. Call me a savior of the friendly fowl. Daniel and I made this recipe as part of our Mar Vista twelve-course meal for twelve. We used Jason Drew's Albariño and Pinot Noir wines, serving his and Molly's wines from Drew Winery in Elk, California, during our entire meal . . . well worth a road trip to their charming winery at 31351 Philo Greenwood Road. Taste the finished pâté with any of the Drew Syrahs— magnificent! You can substitute any three different varietal mushrooms for this earthy, scrumptious appetizer.

Ingredients

2 lb fresh chanterelle mushrooms

½ lb fresh shiitake mushrooms

½ lb Mendocino organic portobello mushrooms (2 large stemless)

¼ c Global Gardens 100% Mission Extra Virgin Olive Oil

½ c Drew Albariño (or sauvignon blanc)

½ c Drew Pinot Noir (Balo Vineyard)

½ c minced shallots

½ tsp cumin

1 ½ tsp fine sea salt

2 egg yolks

1 tbsp freshly minced tarragon

1 stem fresh Italian broadleaf parsley

Method

Chop all mushrooms into ¼-inch pieces and sauté lightly in olive oil with wine, shallots, and spices until mushrooms are just soft, about 10 minutes. Remove from heat and blend in a food processor. Add egg yolks one at a time, then add tarragon. Pat evenly into a 12x4-inch Le Creuset pâté loaf dish and garnish with stem of fresh parsley.

Bake at 350˚, uncovered, for 20–25 minutes until edges are lightly browned. Serve from pâté dish. Cover any remaining leftovers and enjoy within one week.

SPINACH PITA
CALITERRANEAN

MAKES 24 PIECES

Easy, fantastic protein all under one cover that is phyllo, this recipe can be baked and kept refrigerated for 3–4 days. Reheat in a 200° oven for 20 minutes prior to serving. Makes a popular appetizer or main dish.

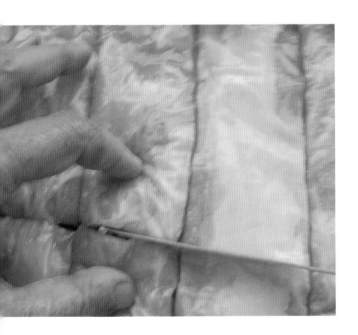

1 medium sweet onion, diced

1 lb fresh, organic spinach (I like the prewashed and trimmed spinach from Earthbound Farms.)

8 oz Greek feta cheese (I prefer FAGE or Mt. Vikos)

1 c kasseri cheese, shredded

½ tsp fine sea salt

⅛ tsp ground black pepper

½ tsp ground nutmeg

1 lb phyllo

1 ¼ c Global Gardens Kalamata or Manzanilla Extra Virgin Olive Oil

Method

Brown onions in 1 tbsp olive oil in a skillet over medium-high heat. Squeeze water thoroughly from spinach in small fistfuls and place into large bowl, stirring in eggs, cheeses, and spices. Add the cooked onions when cool enough not to cook the eggs in the bowl. Lightly oil the bottom of a 13x9-inch baking pan. Coat 20 sheets of phyllo evenly with olive oil. Add spinach mixture and pat down evenly in pan. Coat remaining sheets of phyllo as a top layer over the spinach mixture. Cut through phyllo into 24 equal pieces.

Bake at 375° for 35–40 minutes until evenly browned. Cut pita again to remove from pan.

OLIVE OIL & VINEGAR FOR LIFE

MINI CHÈVRE POPPERS

I always make these quick, astonishingly popular appetizers for large parties. People love them as they pair expertly with white or red wine and are hearty enough to quiet grumbling tummies, yet light enough to enhance the palate for the main course.

Ingredients

- 1 lb log plain chèvre cheese
- 3–4 tbsp Global Gardens Greek Spice Blend
- 40 small leaves butter lettuce
- 1 c Global Gardens Mission/Manzanilla Extra Virgin Olive Oil

Method

Bring cheese to room temperature and form into ¾-inch balls. Coat with spice blend. Place on top of lettuce to form individual serving "cups." Drizzle with olive oil just before serving.

SAVORY BRUSSELS SPROUTS

If you can buy a whole stem of brussels sprouts, do it! They are fun to prepare, and the stalk makes a tasty addition to stir-fry dishes after peeling and sautéing in olive oil and garlic. Look for small brussels sprouts to avoid bitterness. We love to toss this dish with leftover salmon or chicken for a lemony sensation of blended flavors as a main entrée. Crumbled feta or chèvre cheese adds a savory quality.

SERVES 4

Ingredients

1 lb fresh brussels sprouts, washed

½ c purple shallots, thinly sliced and separated

¼ c Global Gardens Lemon or Greek Koroneiki Extra Virgin Olive Oil

1 tsp Mendocino Sea Salt

fresh ground pepper

½ Meyer lemon cut into 4 equal slices

Method

Toss all ingredients together in a bowl, coating brussels sprouts well with the olive oil and salt mixture. Place onto cookie sheet in a single layer.

Roast for 30 minutes at 350°, turning once after 15–18 minutes. Shallots and brussels sprouts will caramelize a bit, perhaps even sticking to your pan—you want that delicious flavor! Garnish with fresh lemon quarters. This dish makes a superb cold salad too.

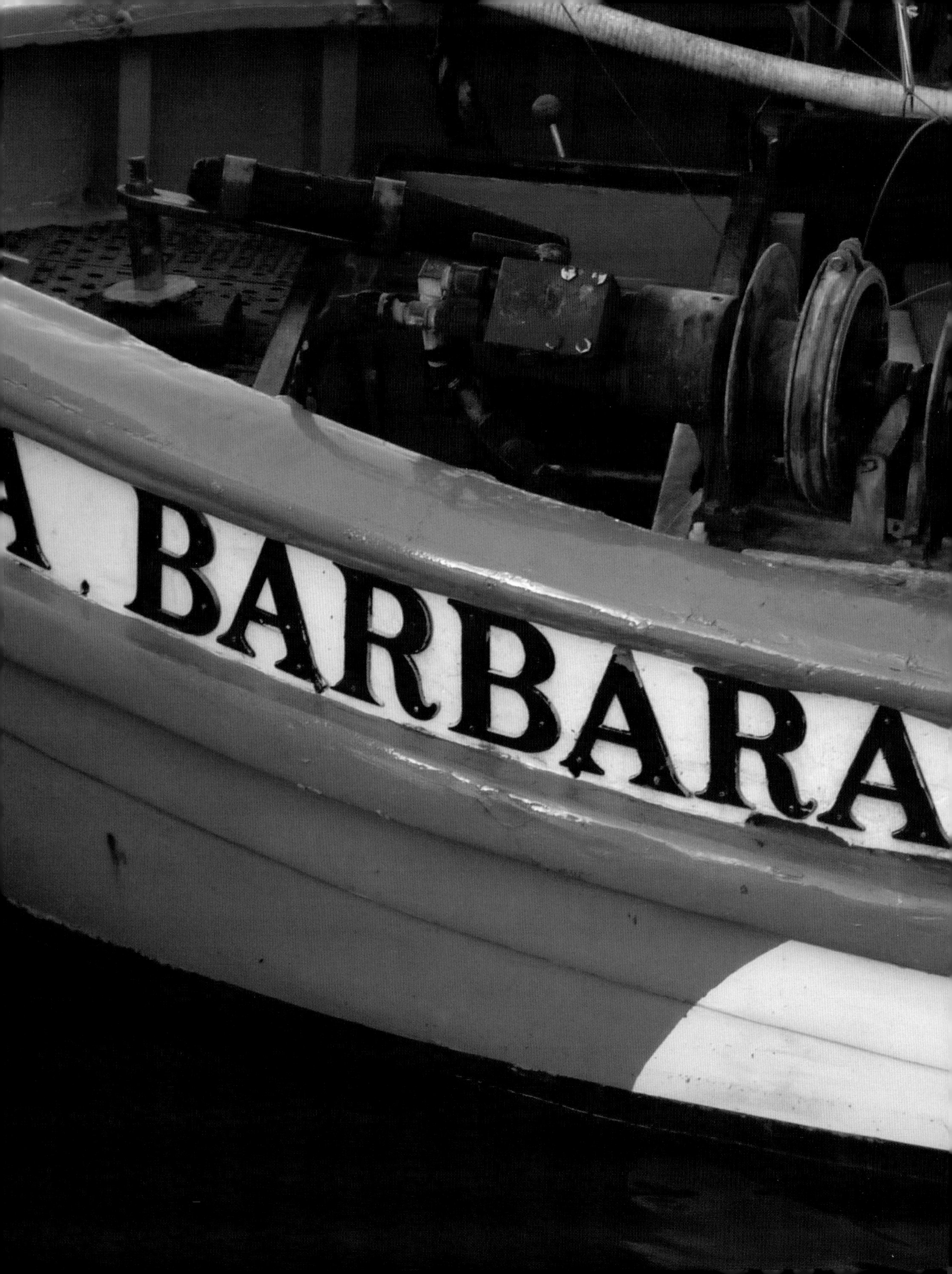
BARBARA

Caliterranean

SEAFOOD

You don't have to live on the coastline of our marvelous country to enjoy fresh seafood. I'm from Dayton, Ohio, and Santa Monica Seafood delivers to Jay's Restaurant and Kitchen Store several times a week. Reputable seafood purveyors are prolific, but please do ask what days of the week deliveries are made, where the seafood is from, and what the freshest available product is at the time of your purchase. Each of these recipes could actually be substituted for whatever fresh seafood you might have; however, cooking times will vary. Lighter fish, for example, will become a lovely ceviche rather quickly as opposed to the 20–30 minutes the scallop ceviche recipe will take. The Szechuan recipe is supreme for scallops and chunks of thick white fish like halibut or monk.

CREDIT: NANCY YAKI

SCALLOPS CEVICHE WITH RED CURRY DELIGHT SAUCE

Sea scallops, to me, are better than lobster! Their texture is perfect for frying, sautés, and . . . ceviche! A lot of people are afraid of ceviche, but fear not! The acid in my vinegars (and in the more traditional Spanish ceviche, which is "cooked" with only lemon or lime) thoroughly cooks seafood without heat or anything else. Their generous size ensures a well-proportioned bite you don't have to struggle for (you buy them already out of the shell). Choose scallops that are even in height, diameter similarity does not matter. Mild yet sweet, they balance nicely with my Red Curry Delight Sauce. Add Cold Heaven Viognier from the Stolpman Vineyard and take your palate on a truly heavenly journey. Any of the Viogniers from Santa Barbara County will do.

Ingredients

6 large sea scallops

1 c Global Gardens Pomegranate Golden Balsamic Vinegar

¼ c organic sour cream

2 tbsp Thai Kitchen Red Curry Paste (found in just about every grocery store)

1 tbsp minced Thai basil leaves

Thai basil leaves for garnish

Method for Ceviche

Place scallops in a bowl that is deep enough to immerse and cover scallops with vinegar. Add vinegar as needed to ensure immersion. Soak scallops for 20–30 minutes, depending on thickness. When scallops are done, they will have a definitive pink color on the outside and be white (not clear or silvery) when thinly sliced.

Method for Sauce

Mix sour cream, curry paste, and minced basil leaves briskly in a bowl.

Presentation

Place sauce into a sandwich-style bag and cut a very tiny portion away from one of the corners of the bag. Squiggle the sauce onto thinly sliced scallops, piled onto individual serving dishes, and garnish with Thai basil leaves.

SZECHUAN PASSION SHRIMP

Szechuan made embarrassingly easy. A true favorite in our household. Make it a main dish by adding chunks of sweet red peppers, fresh broccoli stalks, and fronds of long-cut scallions. Serve over brown rice or noodles.

Ingredients

- 8 wild shrimp
- 2 tbsp Szechuan spices (a ready-made blend found at the store works great!)
- 3 tbsp Global Gardens Mission/ Manzanilla Blend Extra Virgin Olive Oil
- ½ c Global Gardens Passion Balsamic Fruit Vinegar
- ½ c radish sprouts

Method

Peel and devein shrimp. Coat each shrimp generously with spice blend.

Cook in a wrought iron or other heavy clad skillet, warmed on medium-high. Add olive oil and quickly cook shrimp until they just begin to curl, about 3 minutes. Remove shrimp and add the vinegar to the skillet. Bring to a boil, reducing liquids to about 50 percent and a lovely deep bronze color as shown in the recipe photo. Garnish generously with radish sprouts. Did someone say, "*More*, please?"

CRAB 'N' PASTA TO DIVE FOR

OK, I'm lucky, I'll admit it. The Pacific Ocean adorning the Santa Barbara coastline harbors the most flavorful Dungeness crab you've ever tasted. John Wilson has been fishing the waters off our succulent coastline for over twenty years and sells his bountiful catch weekly at the Sunday Hollywood Farmers' Market. If you see him, tell him I sent you. Whatever John features weekly will be the freshest, most flavorful seafood offering you will find anywhere, guaranteed . . . and if his wife Tina is around, ask her for the secret to her succulent crab cakes!

SERVES 2 DECADENTLY, SERVES 4 REASONABLY

Ingredients for Pasta (Thank You, Daniel!)

1 egg

1 c all-purpose flour

⅛ tsp salt

1 tbsp + 1 tsp Global Gardens Farga Extra Virgin Olive Oil

¼ c lime basil, minced

2 tsp water

Method for Pasta

In a food processor, combine all ingredients until you can gather the mixture into a ball.

If you don't have a pasta machine, homemade fettuccine is easy! Flour a pastry board (or your clean kitchen counter). Roll the dough out to 1/16-inch thickness. Cut into long strips with a pizza cutter. Sprinkle with cornmeal and wrap in parchment, refrigerating for up to 6 hours if you are not going to boil the pasta immediately. (But it's always best to make it fresh, if possible.)

Boil 4–5 minutes in 1 quart water containing a splash of olive oil until al dente. Drain and serve immediately with Crab to Dive For topping. (This is probably my favorite main dish in this book!)

- 2 whole Dungeness crabs
- 4 qt water
- 1 tsp salt
- juice of one lime, then cut into chunks for boiling water
- ¼ c shallots
- ⅓ c fresh peas
- 4 tbsp + 2 tsp Global Gardens Farga Extra Virgin Olive Oil
- 2 tbsp fresh savory leaves
- 2 tbsp Global Gardens Pomegranate Balsamic Vinegar

Method

This can be done a day ahead of time and refrigerated, if necessary. Place fresh crab in boiling water containing salt, lime juice, and lime chunks. Boil about 8 minutes per pound (I cooked these two together for 17 minutes). Set your timer! Cook crabs separately if you don't have a big-enough pot. Cool just enough to clean the crabmeat from the shells, about an hour. If you refrigerate the crab in the shell, it will be more difficult to clean.

Sauté shallots in a small skillet with 2 tsp olive oil until lightly browned. Add peas and savory until peas become a bright green. Remove from heat, add 4 tbsp olive oil and pomegranate vinegar. Stir in crab and scoop onto hot pasta.

SEA URCHINS IN PARADISE

OMG, OMG! Fresh sea urchins are my to-die-for seafood. I am immersed in California seafood—or I *thought* I was until I tried urchins. I could literally write a book about them and their delicious buttery quality unsurpassed by their creamy sensual texture. *Uni* (as they are called in Japanese) are mostly exported to Asian countries from California. Santa Barbara county alone harvests about five million pounds of *uni* while Mendocino County brings in about three million pounds.

Tip: A terrific website, calurchin.org, has just about everything you didn't know about urchins, posted along with pictures on how to easily clean them and videos on their sustainable harvest. If you've ever stepped on an urchin while traversing ocean waters, you'd never think that such a spiny, painful species would be the most succulent, flavorful seafood . . . but alas, this is true!

SERVES 4

Ingredients

- 1 small loaf of ciabatta or baguette
- 3 tbsp Global Gardens Farga Extra Virgin Olive Oil
- 2 whole sea urchins, or 8 pieces of processed fresh raw uni
- ½ c Global Gardens Meyer Lemon Balsamic Bliss

Method

Slice bread into 1-inch pieces. Dip one side of bread into olive oil and place onto serving plate. Lay *uni* roe section onto bread. Top with 1 tbsp Bliss and, I promise, the first bite will show you the meaning of utopia!

This is a great way to introduce yourself to urchin, but you can treat them similarly to oysters, making stew, frying them using the Lemon Garlic Veggies recipe on page 157, tucking them into sandwiches and pasta sauces. Superunicaliterraneanexpialidotious!

Tip: You can ask your local seafood market to cut and clean the sea urchins for you, but I think this is the fun part! If you don't live in an area that harvests *uni*, buy them processed from your local seafood market.

OLIVE OIL & VINEGAR FOR LIFE

Caliterranean

FRYING

Enjoy fried foods? *Yes*, you can fry with true, unadulterated extra virgin olive oil, the smoke point is 385°.

Tip: It's practically impossible to regulate the heat in a wok or skillet, so invest in a little deep fryer. I have a Presto FryDaddy Elite. They are not expensive and take up little space in your cupboard. Use Global Gardens Kalamata or Koroneiki Greek Extra Virgin Olive Oils in your fryer. You may use the entire bottle, and remember, you can reuse the extra virgin olive oil in your fryer 4 or 5 times without losing the healthful benefits. Just make sure to scoop out any food remnants after you're done and both the oil and fryer have cooled.

The amazing thing about frying with extra virgin olive oil is the low absorption rate, because it is a monounsaturated fat. Go ahead and measure the remaining oil after you are done frying. You will see that frying for 4–6 people will only use about 4 tbsp of oil. You can use either of these frying recipes on just about any foods you want to fry, such as vegetables, chicken, oysters, scallops, etc. See some suggested substitutions within each individual recipe.

CREDIT: ANITA WILLIAMS

LEMON GARLIC VEGGIES

Who says vegetables are boring? Not these! Caliterranean frying makes these light and tastefully memorable. Fresh lemon and crunchy sea salt are significant contributors to the final succulent quality that will appeal to the masses, bringing shouts for more.

Ingredients

3 tbsp or zest of one large lemon (Meyers are the best, if you can get them!)

1 tbsp or 1 minced clove Global Gardens Elephant Garlic

½ tsp Mendocino Sea Salt crystals

½ c organic crimini mushrooms (have more flavor than standard buttons)

2 small sweet onions, quartered

1 6-inch broccoli crown, separated

8 small okra

½ c all-purpose flour

2 eggs beaten, pinch of salt

1½ to 2 c panko crumbs

1 lemon, cut into six wedges

Process

In a small bowl, stir together the lemon zest, elephant garlic, and sea salt, then set aside. Lightly coat clean veggies first with flour, then egg, then panko.

Fry

Place gently into preheated deep fryer with heat-proof, silicone slotted spoon or spatula. Fry until golden brown, about 2–3 minutes. Remove from fryer and sprinkle generously with the lemon zest, minced garlic, and sea salt mixture. Sprinkle with fresh lemon juice, crunch and munch—great for lunch!

SACRED SHRIMP

Hemp is a super healthy plant with a history dating back some 12,000-plus years. Medeival "gruel" was made of hemp seed meal. Hemp oil is the richest known source of polyunsaturated essential fatty acids (the "good" fats). It is also high in several essential amino acids, including gamma linoleic acid (GLA), a very rare nutrient also found in mother's milk. Hemp seeds are full of this oil and are available in a growing number of products, including my favorite granola, found in the bulk products area of your local grocery store. Using the granola as a coating for frying adds a dimensional flavor character and crunchy texture that's unequalled!

SERVES 2 AS MAIN DISH OR 4 FOR APPETIZERS

Ingredients

1 lb 12–15 ct wild shrimp

¼ c all-purpose or gluten-free flour

1 egg, beaten

2 c hemp seed granola

First

Wild shrimp are typically beheaded and frozen immediately into 5-lb blocks after netting. They are much sweeter and more flavorful than farm-raised shrimp. Here on the West Coast we can frequently get Santa Barbara Channel Island shrimp, which are more in the smaller 28–32 count size ("count" meaning the approximate quantity per pound) but still perfect for this recipe. Freshness matters! Shrimp are easy to peel and devein. Buying them already done is a big flavor loser, so do it yourself. Pinch the tail firmly just above the tail line and before the last break line on the spine. Twist, pinch, and pull simultaneously, and the entire tail meat should be revealed. The remainder of the shell should come off in one piece if you grasp the middle of the inside feathery part of the shell and peel outward. Using a sharp knife, score the spine of the shrimp about ⅛ inch from top to tail, prying apart to show a dark vein that should also come out in one string with the tip of your knife, thumb, and forefinger guiding it.

Method

Lightly coat shrimp first in flour, then egg, then granola.

Fry until golden brown and shrimp are slightly (not too tightly) curled. Do not overcook or the shrimp will be chewy.

CREDIT: ANITA WILLIAMS

OLIVE OIL & VINEGAR FOR LIFE

I usually come home, look in the refrigerator, and start throwing things together, running out to the garden for herbs, Meyer lemons, pomegranates, peaches, kumquats, limes, persimmons, or whatever else is in season either at my place, my neighbors', friends', and at my weekly local farmers' markets. I never measure. Writing this cookbook has shown me that I have to stop, literally, and smell the flowers. Measure, write, taste, cook, taste again, and again. I love to cook even after a crazy long, busy day, and my girls are well accustomed to having snacks after school, when I'll announce our eventual dinnertime. Sometimes 6:00, sometimes 8:30, as long as I get to work with fresh ingredients and new flavors, I'm pretty happy, and they seem to be too. This section provides you with a foundation for cooking an exponentially infinite amount of Caliterranean cuisine recipes. Remember this short list and keep fresh herbs at your fingertips in pots, windowsill gardens, vertical rooftop gardens: fresh, locally available herbs, extra virgin olive oil, and balsamic fruit vinegars.

LEMON CHICKEN
UTOPIA

Bliss, anyone? This dish is so fast and easy to make. My family loves dark chicken meat, but you can use boneless skinless breast tenders instead; sometimes I mix them together.

Ingredients

1.5 lb boneless, skinless organic chicken thighs

½ tsp sea salt

1 tbsp Global Gardens Kalamata Extra Virgin Olive Oil

2 tbsp thinly sliced Global Gardens Elephant Garlic

1 c Global Gardens Meyer Lemon Bliss Balsamic Vinegar

Wash chicken and pat dry. Cover each side with ¼ tsp sea salt. In a wrought iron skillet over medium heat, add the oil and fry the garlic slices for 2–3 minutes just until the edges begin to brown, making sure to coat them evenly with oil.

Tip:

Don't stir them around too much or they won't brown, let them fry for about 1.5 minutes before stirring.

Move browned garlic slices to the edges of the skillet. Pour Bliss into hot skillet and arrange chicken evenly. Cook without covering, about 10 minutes; the liquid will boil. Turn chicken and cook the other side 10 minutes. Turn heat to medium-high and turn chicken again for 3 minutes until the liquid thickens and the meat is brown. Enjoy in utopian pleasure.

THAI WONDER

We love the influence of Thai flavors, and they are so easy to create at home. This recipe takes all of 20 minutes to prep and about 10 minutes to cook. Better and faster than ordering Thai food for home delivery. The dressing recipe is also great for seafood, chicken, and tofu. Sometimes I even use leftover fish or chicken to create this salad.

SERVES 4

Ingredients for the Salad

1 lb Diestel Ground Turkey (great tasting, no hormones, vegetarian fed)

1 tbsp Global Gardens Farga Extra Virgin Olive Oil

1 tbsp Thai Kitchen Red Curry Paste

1 bunch (1 c) green onions (include green tops)

2 c green cabbage

1 c purple cabbage

½ c diced fresh carrot

8 oz rice noodles

3 tbsp minced mint

lime quarters

Ingredients for the Dressing

3 tbsp green Thai Kitchen Curry Paste

3 tbsp organic peanut butter

3 tbsp fresh lime juice

¼ c + 1 tbsp water

Method

Cook turkey in a skillet with olive oil and red curry paste. Remove from heat. Slice green onions and cabbage, then toss with carrots. Boil noodles for 5 minutes or according to package instructions;

do not overboil or they will quickly get mushy!

Mix dressing ingredients with an immersion blender or small food processor.

Presentation

Place vegetable mixture onto individual serving plates over boiled noodles. Add

a cluster of ground turkey mixture with an ice cream scoop. Drizzle dressing and garnish with fresh mint and lime quarters. Encourage a fresh squeeze of lime juice from participating dining guests!

POMEGRANATE BABY BACK RIBS & SHROOMS

SERVES 2

Mar Vista Cottages are owned by Tom and Renata Dorn of Anchor Bay, California. Hands down my favorite place to relax real time on California's most scenic Highway 1, Mar Vista offers a phenomenal organic garden stocked to the deer fence line with seasonal root and aboveground vegetables, herbs, berries, beans, and flowers of the edible and table variety to please all senses and sensible people. Anita, Sunita, and I found Mar Vista quite by accident many years ago. Nostalgic, spotless cabins with wood floors, a properly equipped kitchen and fresh hen eggs at your dawn-swept door in the morning are all commonplace here. One can't help but mingle with the chickens and goats, listening to nasturtium flowers talking to the beets and fava bean flowers. Pure joy greets you at every smiling turn. The next five recipes were conceived here along with several others scattered throughout the book, most notably Mar Vista Crème Brûlée (see page 220), and Smoked Salt Chocolate Tart (see page 219).

Ingredients

- 1 rack antibiotic-free, natural pork baby back ribs
- 1 tbsp + 2 tsp Global Gardens Santa Ynez Italian Varietal Blend Extra Virgin Olive Oil
- 2 tsp Mendocino Sea Salt Crystals
- 6 oz Pomegranate Balsamic Vinegar
- 3 cloves Global Gardens Organic Elephant Garlic
- 4 king trumpet mushrooms

Method

Brush ribs with olive oil and lay them, meat side down, in a rectangular casserole dish. Sprinkle bone side of ribs with sea salt. Pour vinegar into the casserole dish. Make sure vinegar covers the depth of the meat to the bone within the dish. If not, add vinegar to reach meat height. Mince garlic into the dish. Marinate for 8–12 hours in the refrigerator.

Bake at 225˚ for 1 hour; meat and vinegar juices will reduce naturally. Place ribs onto a hot grill. Coat raw king trumpet mushrooms first with 1 tsp of olive oil each, then heavily with reduction juices. Put mushrooms on grill. Turn both ribs and mushrooms when grill marks are prevalent.

OLIVE OIL & VINEGAR FOR LIFE

CREDIT: RON BOLANDER

SEAWEED POWER PASTA

Textures, colors, and flavors unite! The powerful, healthful quality of palm seaweed combined with the ease of boiling pasta, opening a package of Donna Bishop's hand-harvested goodness along with protein from Global Gardens and a reduction glaze made popular in this book makes a most delicious seaside dinner.

SERVES 4

Ingredients

1 c Global Gardens Fig Balsamic Vinegar

1.75-oz pkg Donna Bishop's dried sea palm

1-lb pkg organic semolina noodles, farfalle or penne

3 tbsp Global Gardens 100% Mission Extra Virgin Olive Oil

5-oz tin Global Gardens Organic Spicy Walnuts

Method

Reduce vinegar by 50 percent, as in the Balsamic Reduction recipe on page 27.

Rehydrate dried sea palm for a few minutes in hot water. Boil pasta according to package directions (or make homemade pasta from the Crab 'n' Pasta to Dive For recipe on page 148). Use 1 tbsp olive oil in boiling water to prevent pasta from sticking when draining. Drain al dente pasta in colander with sea palm and immediately transfer to large serving bowl and toss together with remaining olive oil and spicy walnuts. Make creative swirls over entire pasta dish with reduction glaze.

SMOKED MENDO TRI-TIP

I never had a beef tri-tip cut prior to moving to California, where it is quite common at BBQs and on steakhouse menus. The tri-tip is a cut from the bottom sirloin primal, so there are only two per animal. The meat is chewy but quite flavorful. I made this meal the day I met Lora La Mar at the Gualala Farmers' Market. The exact flavors are impossible without Lora and Bob's Sea Smoke, handcrafted artisan sea salt that's slow-smoked over sea kelp, maple, and alder woods.

Tip: You can make this with a frozen tri-tip taken out of the freezer the night before eating and following the method on the right.

Ingredients

1 ½–2 lb tri-tip

2 tbsp Global Gardens 100% Mission Extra Virgin Olive Oil
¼ c Global Gardens Fig Balsamic Vinegar

1 tbsp Sea Smoke

Method

Rub tri-tip with olive oil, then place into a Ziploc bag. Add vinegar and sea salt and seal the bag. Marinate all day, minimum 8 hours, or up to 24 hours.

Grill the tri-tip on hot coals, turning as each side gets grill marks, about 30 minutes total. Charred areas will be well-done, with rare pieces sliced from the middle of the tri-tip.

OLIVE OIL & VINEGAR FOR LIFE

CHICKEN PINK SALAD

SERVES 6

I'll admit I've had a love affair with beets ever since I was a child and my grandmother told me to stop eating so many or my insides would turn into one (which they did!). But I've never had the pleasure of grating raw beets into a salad and watching everything turn the most magical, naturally edible color . . . *and the flavor*! Well, this is a must-try recipe. Don't let oven-roasting a chicken get the best of you. Go ahead and buy a premade oven-roasted chicken if you must, or use the Oven-Roasted Chicken Supreme recipe on page 185.

Ingredients for Meat Mixture

deboned meat from one whole organic or free-range chicken, finely chopped

⅓ c green onions with tops

⅓ c celery

⅓ c raw organic walnuts

⅓ c fresh raw peas

⅓ c shredded red beets

½ tsp fine sea salt

½ tsp fresh ground mixed peppercorns

Ingredients for Mendo's Green Goddess Dressing

½ c fresh parsley

½ c fresh tarragon

½ c fresh chives

1 clove Global Gardens Elephant Garlic

1 ⅓ c Global Gardens Mission Olive Oil

Juice of 2 limes

Method

Toss all meat mixture items into a large bowl. Use an immersion blender to mix dressing ingredients. Stir dressing into chicken mixture and serve.

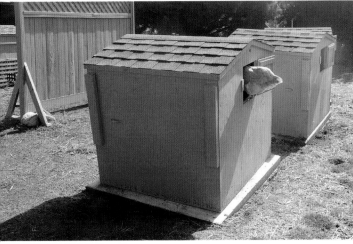

PETALUMA LAMB STEW

We were disappointed to get to Mendocino County in late October and find that the fishing season had ended—the only thing available were sea urchins! The results of that realization, however, are two far superior recipes to the Mendocino Seafood Stew we *were* going to make. If you're reading chronologically, you've already discovered the delicious recipe for Sea Urchins in Paradise. Our second new recipe was made with local, naturally raised lamb as the foundation of this decadent dish. I have to hand it to Daniel, thinking on a whim and a lamb leg. I was stuck on making the seafood stew. A few short hours later, I was a humble lover of this incredible *Petaluma Lamb Stew*!

SERVES 4

Ingredients

- ¼ c + 3 tbsp Global Gardens Farga Extra Virgin Olive Oil
- 3 whole lamb shanks
- ¼ tsp salt
- freshly ground pepper
- ½ c Pinot Noir (we used Drew Winery Fog Eater)
- 1 medium clove Global Gardens Elephant Garlic
- ½ c diced shallots
- 2 whole large carrots, peeled
- 2 stalks celery, peeled
- ¼ c Italian broadleaf parsley leaves
- 3 sprigs fresh thyme, tied with kitchen string
- pinch of fine sea salt
- 2 c fresh tomatoes, chopped (if in season; if not, I like Muir Glen organic canned tomatoes)
- 1 tbsp cumin
- 4 slices pancetta

Method

Lightly coat a medium large roasting pan with about 2 tsp olive oil. Coat all sides of shanks with salt and pepper and remaining olive oil. Roast in a 350°

oven, uncovered, for 50–55 minutes or until nicely browned. Remove from oven and place on a warmed dish, cover with foil to keep warm. Pour off any fat from roasting pan and deglaze with wine.

In a small skillet, sauté garlic and shallots in 2 tbsp olive oil. Pour deglazed liquids and bits into a deep skillet or dutch oven large enough to hold remaining ingredients. Add lamb shanks and garlic mixture, along with 1 carrot, celery, parsley, and thyme bunch. Simmer on low for 1½ to 2 hours until lamb is almost tender. Quickly sauté last carrot, sliced diagonally, in 1 tbsp olive oil, seasoning with a pinch of fine sea salt. Add sautéed carrots and tomatoes to stew pot.

In the same skillet you used to sauté the carrot, fry pancetta slices until crispy. Simmer for 20–30 more minutes. Remove whole carrot, celery, and thyme bunch, discard. Remove lamb shanks and slice in portions to serve. Garnish with pancetta slices.

CREDIT: RON BOLANDER

OLIVE OIL & VINEGAR FOR LIFE

CITRUSY SKIRT STEAK

You won't believe this less-than-ten-minute entrée. Moist enough to cut with a fork, zesty, and original in flavor. Everybody loves this dish with oven-roasted corn on the cob (see the Cowgirl Blue Caviar recipe on page 115) and my Avocado Olive Salad (see page 131).

Ingredients

- 1 c Global Gardens Blood Orange Balsamic Vinegar
- 1½ lb antibiotic- and hormone-free skirt steak
- 2 tsp Caliterranean Garden Blend
- 1 blood orange, sliced

Method

Pour the vinegar into a cast iron or other heavy skillet over medium-high heat and bring to a boil. Once the vinegar begins to bubble, carefully place the sliced steak into the skillet, turning it quickly to keep the meat medium-rare, the optimum way to serve and eat this dish. It will cook in less than a minute! Remove the steak from the skillet and reduce the remaining juices at a medium boil for a few minutes. When it's thick enough, drizzle the reduced warm juices over the meat. Garnish with orange slices.

STUFFED SWEET PEPPERS & ZUCCHINI

This is a healthier, more flavorful twist to traditional Greek *yemista*. I make the dish using fresh peppers from the garden all year long.

Tip: You can freeze whole, unwashed sweet peppers in a Ziploc bag for use all year long! It's one of those recipes you can make one day and eat for the rest of the week—and the flavors blend together more beautifully by the second or third day.

I use the sugar inside the veggies because sometimes oven-roasting peppers can make them a little bitter. The amount of sugar is negligible from a dietary standpoint and makes a nice difference in the final outcome of the recipe.

SERVES 6

Ingredients

- 6 assorted red and yellow peppers with tops sliced off and seeded
- 2 2-inch zucchinis with top and bottom removed, sliced in half and scooped out, leaving 1 inch on the bottom
- 2 small potatoes, quartered
- 2 tsp organic sugar
- 1 ½ c brown rice (more fiber and less carbs)
- 1 large purple shallot
- 1 clove Global Gardens Organic Elephant Garlic
- ¼ cup + 3 tbsp Global Gardens Farga Extra Virgin Olive Oil
- 28-oz can Muir Glen Fire Roasted Whole Tomatoes
- ¾ tsp salt
- ¼ tsp fresh ground pepper
- 1 15-oz can organic black beans (adds 6 g protein per serving over other beans!)

Method

Pack peppers and zucchini tightly in a 12-inch baking dish or roasting pan. Place potato quarters around edges of the pan, in between veggies. Dust the insides of peppers with 1 tsp sugar. Keep lids of peppers for oven roasting when dish is completed. Brown the rice, shallots, and garlic in 3 tbsp olive oil until rice is somewhat clear, about 5 minutes in a dutch oven or equally sized heavy pot. Add tomatoes, salt, and pepper, bringing mixture to a boil over medium-high heat for about 3 minutes, stirring frequently so that the rice does not stick to the bottom. Reduce to a simmer, add beans, and stir gently so as not to break them, simmering for another 3–4 minutes. Remove pot from heat. Use a slotted spoon to insert mixture into open vegetables, about halfway up their sides. Add any remaining mixture in between the open veggies, at the bottom of the baking dish or pan. Fill veggies to the top with liquid from the pot; this will give the rice liquid to absorb, expanding and filling the peppers and zucchini in the baking process. If you run out of liquids, use water to fill the veggies. If you tuck uncooked rice mixture in between veggies, pour enough water over those areas for rice to absorb. Drizzle open veggies with ¼ cup olive oil and dust with remaining 1 tsp sugar. Put lids on peppers.

Cook

Tightly wrap dish or pan with aluminum foil so steam will not escape. Cook at 500° for one hour. Slide oven rack out and carefully lift foil so as not to burn yourself with the steam. Do a taste test with the rice to make sure it is soft and cooked thoroughly prior to removing from oven.

EGGPLANT CALITERRANEANNA

Love the man who will happily make you eggplant! (Thank you, Daniel.) This recipe looks complicated because of the list of ingredients, but, honestly, it's not. The flavors, like so many oven-baked dishes in this cookbook, are better on the second day after cooking. If you know you're going to have a tough week, this is a fun recipe to make on a Sunday night and pairs wonderfully with Qupé Marsanne.

Tip: Make the sauce first. Either double the recipe to use all the sauce, or use leftover sauce for pasta or pizza during the same week. You can also freeze any remaining sauce in a freezer container for later use.

SERVES 4

Ingredients for Tomato Sauce

¼ tsp red pepper flakes

¾ c chopped onion

⅛ c Extra Virgin Olive Oil

3 cloves garlic, chopped

¼ c carrot, peeled and diced

¼ large red pepper, chopped

1 tbsp fresh thyme, chopped

⅓ c pancetta, finely diced.

½ c dry white wine

1 28-oz can Muir Glen organic chopped tomatoes

2 tbsp fresh basil, chopped

¼ tsp fine sea salt

¼ tsp freshly ground pepper

Method

Sauté pepper flakes and onion in olive oil over medium heat, 2–3 minutes. Add garlic and carrot for another 2–3 minutes. Add red pepper, thyme, and pancetta and cook for another 5 minutes. Add wine; reduce until about only 1 tbsp of liquid is in skillet. Add tomatoes. Simmer until the mixture reaches the desired consistency—like a very thick soup.

Ingredients

1 large or 2 small eggplants, sliced ½ inch thick. Try to cut to shape of the oven dish(es) to be used in final assembly

¼ c Global Gardens Farga Extra Virgin Olive Oil

⅛ tsp fine sea salt

freshly ground pepper

1–1¼ c tomato sauce (see recipe to the left)

6 oz buffalo mozzarella, sliced into ¼-inch pieces

¼ c freshly grated Parmesan cheese

3 tbsp fresh basil

1 tbsp panko bread crumbs

Method

Place eggplant slices in a single layer on a large cooking sheet. Spread olive oil, salt, and pepper over eggplant slices.

Bake at 450˚ for 14–15 minutes.

Reduce heat to 350°.

In a 6x9-inch baking dish, spread ⅓ of the tomato sauce. Place ½ of the eggplant, in a single layer, over the sauce. Cover with 3 oz mozzarella, ⅛ c Parmesan cheese, and 1 tbsp basil. Repeat. Drizzle top with 1 tbsp olive oil and 1 tbsp chopped basil and panko crumbs.

Bake until top is golden, about 20–25 minutes.

CONFETTI STUFFED PORK TENDERLOIN

Pork tenderloin is one of the leanest cuts of pork available and needs to be cooked quickly on high heat. I make sure I buy all-natural, hormone- and antibiotic-free pork. Tenderloins typically come two to a package, so I always cook both at the same time.

Tip: Leftovers make great sandwiches sliced in half. The veggies in the stuffing make for a nice crunch and add a savory quality.

Tip 2: This recipe makes a great main appetizer feature for cocktail parties. I catered a party in Palm Springs where Martha Stewart was the featured guest. She readily enjoyed these flavorful bites along with her staff as she proclaimed "good things" about Global Gardens food products!

Ingredients for Meat

1 pkg (about 2 lb) pork tenderloin

12 stalks thick asparagus

½ c diced sweet yellow or red pepper

1 clove (1 tbsp) Global Gardens Organic Elephant Garlic, minced

2 tbsp Global Gardens Santa Ynez Italian Varietal Blend Extra Virgin Olive Oil

1 tsp coarse sea salt

Ingredients for Garnish

1 c Global Gardens Passion Fruit Balsamic Vinegar

1 large carrot, finely shredded

24 fresh peas

24 2-inch purple cabbage triangles

Method

Remove clear membrane, if prevalent, from length of meat. Gently pound down meat to ¼-inch thickness. Chop asparagus, pepper, and garlic. Rub each piece of meat on both sides with olive oil. Make a row of chopped asparagus, peppers, and garlic in the middle of the meat. Roll the width of the pork tenderloin and tie in 4 or 5 places down the length to keep it from falling apart

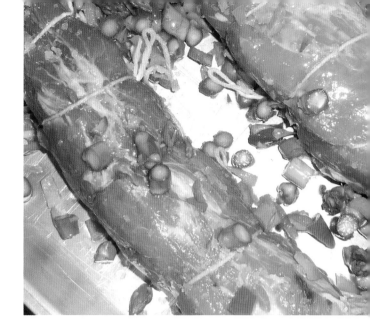

when cooking. Sprinkle both rolls with
sea salt.

Cook

Tightly wrap pan and place in a 500°
oven for 25 minutes. Meat should be very
slightly pink but not brown when sliced
in the middle. Do not overcook! The meat
will continue to cook a bit when slicing
and presenting. Slice meat into ¾-inch
slices, about 12 per roll.

Garnish

Make reduction (see page 27) with
vinegar. Put 1 tsp of reduction onto the
top of each slice. Make a small "nest" as
shown in photograph with the carrots
and place a single green pea in the
middle. Put entire slice with garnish onto
a purple cabbage triangle and serve.

OVEN-ROASTED CHICKEN SUPREME

This is another recipe I feel a little guilty about including. I mean, *it's just roasted chicken, people*! Still, I get customers and friends asking me all the time how to make a great chicken. Simple as it is, here is my favorite recipe. Use it for chicken salad, chicken sandwiches, chicken soup, and chicken crepes—I guess I use it for everything requiring roasted chicken! And remember, chicken roasts from the bone outward, so most of the salt will permeate the bone first before reaching the juices as the meat heats and cooks. I have made this chicken recipe hundreds of times—trust me, it's easy and delicious. Note: the end result may *look* a bit dry, but I promise this will be the juiciest roasted chicken you've ever had.

Ingredients

2 tbsp fine sea salt

1 4.5 lb (average) whole organic or free-range chicken

3 cloves Global Gardens Organic Elephant Garlic, halved

3 whole rosemary sprigs

3 tbsp Global Gardens Ascolano or Farga Extra Virgin Olive Oil

Method

Pour sea salt into your hand and rub the inside of the chicken cavity, getting into as many crevices as possible. Place garlic cloves and 1 rosemary sprig into cavity. Pour olive oil, 1 tbsp at a time, into your hand and, gently lifting the skin away from the breast meat, rub oil mostly under and use a bit of remaining oil over skin, reaching under skin into leg area as much as possible. Insert remaining rosemary sprigs under skin wherever possible.

Roast at 375˚ for 1¼ hours until meat thermometer reads 165˚ on chicken breast, being careful not to touch the bone when inserting. Chicken will still cook a little bit after removing it from the oven, ideally reaching an internal temperature of 170˚ on the breast meat. Let sit for about 10 minutes, uncovered, until cool enough to cut into portions.

GINGERMANIA TOFU

I have a lot of vegan friends and get a lot of pressure to convert. I love vegan food. I just can't bring myself to give up chicken, beef, pork, and seafood every once in a while to round things out. Still, we cook vegetarian meals at home about three nights a week. The trick with tofu, I've found, is to cook it in a skillet separately first, browning it in olive oil with its own spices, creating a savory flavor and a nice texture.

Ingredients

18 spears organic asparagus

24 crimini mushrooms

¼ c Global Gardens Ascolano Extra Virgin Olive Oil

1 tsp minced Global Gardens Organic Elephant Garlic

2 tbsp shaved fresh ginger

1 tsp coarse sea salt

3 tsp masala spice (3 tsp = 1 tbsp, but each tsp will be used separately, so use a tsp)

12-oz pkg organic smooth tofu, sliced into 3/16-inch pieces widthwise and halved lengthwise

1 c Global Gardens Apple Ginger Balsamic Vinegar

6 organic cherry tomatoes, halved

½ c microgreens

Method

In a skillet over medium heat, sauté asparagus and mushrooms with olive oil, garlic, ginger, ½ tsp sea salt, and 1 tsp masala, turning frequently. Remove veggies from skillet and place tofu that has been patted dry and covered with remaining 2 tsp masala and remaining ½ tsp salt. Brown both sides evenly and remove from skillet. Skillet should now be fairly dry. If not, wipe clean with paper towel, leaving bits of tofu or veggies existing.

Pour 1 cup apple ginger vinegar into skillet and bring to a boil. When the vinegar begins to foam, after 3–4 minutes, it is thickening. Add tofu and veggie mixture, coating with the vinegar mixture. Remove from heat when vinegar mixture has browned nicely and has a rich flavor, about 2 more minutes. Garnish with freshly halved tomatoes and microgreens.

Caliterranean

CAMPSITE DISHES

Camping is an integral part of the Caliterranean lifestyle. It is the most convenient and superior way to get unplugged and reconnect with family and friends, while enjoying nature and her lovely offerings. My favorite part of tent camping in California is making sure everyone has happy tummies—being in the elements all day makes you practically ravenous! How wonderful that each state in America has a parks system where we can all easily access some Caliterranean vibes, even if it's just for one night. California has 278 state parks and 23 national parks, each offering distinction and priceless memories. The recipes created in this section were all photographed on site in Death Valley National Park—a thrilling, dramatic landscape with expansive vistas . . . the perfect place to work up an appetite for simple, tasty, and restorative recipes.

I used the same varietal Global Gardens Extra Virgin Olive Oil (Manzanilla) and Balsamic Vinegar (Fig) for lighter traveling. If you're making these dishes at home, these flavor profiles are still perfect, but feel free to be more adventurous.

Tip: the only cooking utensils you really need are a wrought iron skillet, paring knife, bread knife, wooden mixing spoon, plastic 2-cup measuring container, measuring spoons, heavy-duty metal spatula, 4 metal camping bowls, and BBQ tongs—quick and easy, right? Make vinegar reduction ahead of time to use for Master Veggies and Citrus Relief Salad if you don't want to make it on site.

FRESH³

A raw sauce that is an easy concoction, creating a flavorful, healthy main act for three different types of favorite campout dinners—pasta, bruschetta, and salsa—don't forget your favorite dry noodles, bread, or chips to best enjoy this fresh, year-round dish. (Tomatoes sold on the vine work great during the off-season.)

Ingredients

- ½ c chopped mushrooms
- ½ c zucchini
- ½ c sweet yellow pepper
- ¼ c packed minced fresh basil
- 3 tbsp Global Gardens Fig Balsamic Vinegar
- 2 tbsp minced fresh dill
- 1 jumbo clove
- 1 c chopped tomato

Method

Marinate chopped mushrooms for at least 1 hour (can be done up to 24 hours ahead of time and stored in a plastic container). Toss in remaining fresh ingredients.

Tip

Note that the recipe does not call for salt. Salt dehydrates our energy, especially in sunny, windy conditions, and really is not necessary for this flavorful blend of raw ingredients. Salt will also break down the tomatoes quickly. If you do choose to add it, use really firm tomatoes and know that Fresh³ won't taste so great if not eaten immediately. Saltless, you can keep it for up to 24 hours to use on multiple meals—no need to refrigerate! Just keep out of direct sunlight and heat. If you do put it in the cooler (we did in Death Valley), put it on top of everything and not directly on ice. Tomatoes lose their flavor when refrigerated.

BEYOND BRUSCHETTA

Ingredients

1 12-inch loaf French bread or baguette

¼ c Global Gardens Manzanilla Extra Virgin Olive Oil

2 tbsp Global Gardens Fig Balsamic Vinegar

4 oz hard cheese, like Asiago or kasseri, sliced into strips

Method

Slice bread into 1-inch-thick pieces. Dip bread thoroughly (but without soaking) into olive oil on one side and vinegar on the other. Toast bread with olive oil side down first on open fire or grill. Turn bread over on grill, add a generous spoonful of Fresh[3] and a cheese strip on each slice of bread and cover with the lid of a skillet (the skillet itself if you don't have a lid, or other metal camping bowl so that cheese will melt).

MASTER VEGGIES

I have always loved char-grilled vegetables: peppers, artichokes, potatoes, onions, asparagus, eggplant—you can't name a vegetable I have not grilled with extra virgin olive oil and not fallen madly into ecstatic, sudden joy with the first bite.

Tip: I find a better concentration of flavors in smaller-sized veggies, especially potatoes and sweet peppers! This recipe is perfect for any outdoor picnic, using any veggies, even made ahead and eaten at room temperature. The potato adds the necessary carbohydrates for a high-energy day outdoors.

SERVES 4

Ingredients

- 2 3-inch artichokes, boiled just until fork can be pierced into heart (can be done ahead of time at home)

- 2 3-inch potatoes (prebaked at home until a fork just goes through it, not too mushy. They may be boiled if done on site)

- 6 assorted small sweet red peppers, quartered

- ¼ tsp Mendocino Sea Salt

- 3 tbsp Global Gardens Manzanilla Extra Virgin Olive Oil

- ½ c Global Gardens Fig Balsamic Vinegar

- 2 tbsp plain yogurt (I like FAGE 2%)

Method

Cut artichokes and potatoes in half lengthwise, after preparation suggested above. Rub all sides of veggies with sea salt and olive oil. Make reduction sauce by boiling vinegar for about 3 minutes in skillet (you can make this ahead of time and bring along in a small jar or squeeze bottle), cool quickly in cooler before stirring in yogurt.

Grill vegetables about 10 minutes per side over hot coals or campfire grill. Garnish generously with reduction glaze blended with yogurt.

CITRUS RELIEF SALAD

Refreshing, vibrant, filling, and healthfully energetic—easily the perfect summer lunch for a park, the beach, or your own backyard. Serve with Easy Protein Crackers (see page 119) for extra protein and carbohydrate energy.

Ingredients

2 tangerines *or* 1 grapefruit, sectioned

1 c fresh sunflower sprouts

4 whole fresh peas in the pod

¼ c feta cheese crumbles (chèvre works well too)

¼ c Global Gardens Fig Balsamic Vinegar, reduced to ⅛ c

fresh clover flowers (from foraging!) for garnish

Method

Arrange fruit and vegetables in an artful way, opening pea pods to expose peas. Sprinkle with crumbled cheese and drizzle generously with fig vinegar reduction. See reduction recipe from Master Veggies on page 195. Garnish with fresh and zingy clover flowers.

OLIVE OIL & VINEGAR FOR LIFE

DEATH VALLEY DEVILED EGGS

High protein, easy, and flavorful, these mayo-free, healthy deviled eggs are a real treat that will last a couple of days in your cooler . . . if they don't get eaten first! Boil the eggs before you leave home and place them into their original egg crate for convenient, breakage-free transport to your campsite.

MAKES 24 HALVES

Ingredients

1 dozen organic free-range eggs, boiled

⅓ c Global Gardens Indian Curry Mustard (or other mustard of choice)

⅓ c Global Gardens Mission/Manzanilla Extra Virgin Olive Oil

¼ c packed minced cilantro

½ tsp salt

Method

Cut boiled eggs carefully in half. Remove yolks and whisk in bowl with mustard, extra virgin olive oil, cilantro, and salt. These high-protein treats make great protein additions to Master Veggies and Fresh[3] meals.

TRAIL MUSCLE MIX

Protein, fiber, energy, no need for anything else! Fans of raw snacks can make this flavorful, savory snack by blending all ingredients without roasting, allowing it to sit overnight in a plastic container and shaking frequently.

Ingredients

1 c organic raw almonds

1 c organic raw walnuts

1 c organic raw pistachios

2 tbsp Global Gardens Greek Spice Blend

2 tbsp Global Gardens Manzanilla Extra Virgin Olive Oil

1 tbsp Global Gardens Fig Balsamic Vinegar

1 c dried fruit mixture like cranberries, white raisins, apricots (pick your favorite—dried blueberries are awesome too)

Method

Over open fire or on a camp stove, stir all ingredients except fruit in a wrought iron skillet, almonds first for 5 minutes (they take the longest to roast), prior to adding walnuts for another 5 minutes, then the pistachios for the last 5–6 minutes—about 15 minutes total—until nuts are roasted to a deeper hue than their original color. Time required will vary depending on heat source. Stir frequently with a wooden spoon to make sure one side doesn't burn. Typically, as soon as you smell a nutty scent they're done and will continue to cook in the hot skillet—remove them from the skillet immediately. Add the fruit to the nut mixture.

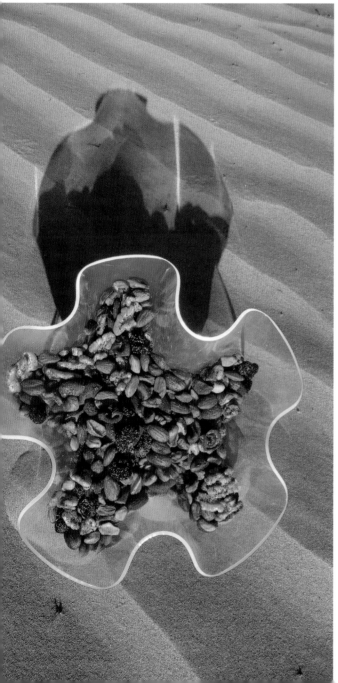 

SOLAR SAUCE

This one is really fun for small children and a great way to educate them about the many ways we can use the sun's energy! It's sweet and tasty to make on a camping trip or picnic—or even in your own special sunny spot at home. The same recipe (different method) is featured as a cooked sauce on page 213 with my California Tangerine Dream.

page 213

Ingredients

1 c 60% cacao chocolate chips (we love Ghirardelli)

⅓ c Manzanilla Extra Virgin Olive Oil

Method

Place chocolate chips into a one-gallon Ziploc-style freezer bag, together with olive oil, and place bag flat in a warm sunny area for a few hours. The fun part comes in squishing it all together as it melts! (WARNING: a quart-sized bag can be used but may pop open while squishing and create a messy chocolate hazard if not done gently.) Snip ⅛-inch hole in one corner of the bottom of the bag to squiggle out when ready.

Tip: Just like solar-operated electronics that don't work without sunshine, Solar Sauce will harden if not kept warm. Either use immediately or reheat on open campfire in a metal bowl, floating on a pot or skillet of water.

CAMPFIRE S'MORES

You *can* roast marshmallows to add to this berry zesty, easy campsite recipe, but if you're anything like me, you'll probably forget to bring along the marshmallows— I'm always too focused on packing just the right olive oil, vinegar, and . . . chocolate!

Ingredients

16 organic graham crackers (We *love* vegan organic Amaranth Graham Crackers by Health Valley)

1 c fresh raspberries

½ c Solar Sauce

Method

Squish fresh raspberries onto crackers and drizzle with Solar Sauce. Look, mom, no fire required!

SOLAR SAUCE SNAX

Ingredients

16 organic graham crackers

1 c sliced organic strawberries

Method

Immerse crackers completely into Solar Sauce and let harden on a plate. Garnish with strawberry slices and go make s'more! Both recipes are wonderfully interchangeable with each other and different fresh fruits, such as bananas or blueberries.

Incredible vistas, magical, mystical Death Valley National Park, Caliterranean camping at its most enjoyable with my dear friend Nancy Garvin

OLIVE OIL & VINEGAR FOR LIFE

Caliterranean

DESSERTS

My mom loved butter. I loved the smell of it as she clarified it—the way her mother had taught her to do. Well, now I love using extra virgin olive oil in place of butter *for everything*. I never thought it could work with chocolate or caramelizing. It does, and I'm a dedicated fan . . . for life!

Inspired by sun, sand, and sea salt, baking desserts with Global Gardens Extra Virgin Olive Oils and Vinegars has never been more satiating—and the Palm Springs desert in the middle of winter (Thank you Dori Bass for your friendship and lovely condo!) was the perfect place for me to perfect my recipes. (I always confuse the spelling—do we vacation in the *dessert* or the *desert*?) Desserts made with extra virgin olive oil in place of butter have a perfect moist quality—the pie crusts are more flaky, and the best part is that after you're finished with a piece, slice, or chunk, you don't feel like a stuffed animal. Sometimes, a great little slice of dessert substitutes perfectly for breakfast on the go, without the guilt—the way our bodies process olive oil helps to make carbohydrate sluggishness history.

NOT YOUR MOTHER'S BAKLAVA

YIELD: 44 PIECES

I was raised eating Greek pastries, gathering with friends and family several times a month and relishing the sheer joy of dripping, gooey, sugary, delectable desserts. My friends and I love this nontraditional, less sugary, but total Caliterranean goodness . . . and for a dessert, goodness is a very sweet thing. Don't be intimidated to try working with phyllo. It's a very rewarding process.

Tip: Remove phyllo from freezer and let stand in refrigerator for 24 hours before using. Let it come to room temperature *before opening box and inside wrapper* for 1 hour before using. If you ruin a sheet because your fingers are too moist or it gets cracked, throw it out. If a sheet tears when oiling it, don't worry about it, just cover with another sheet and move on to the finish line!

CREDIT: ANITA WILLIAMS

Ingredients for the Pastry

- 1 c each raw pistachios, walnuts, and almonds, chopped finely (but not into a paste)
- 3 tbsp organic sugar
- 3 tbsp cinnamon
- 1-lb pkg phyllo
- ½ c Global Gardens Meyer Lemon Extra Virgin Olive Oil
- ½ c Global Gardens Greek Kalamata *or* Koroneiki Extra Virgin Olive Oil

Ingredients for the Syrup

- 1 c organic sugar
- 1 c water
- juice of one lemon (Meyer if available)

Method for Syrup

Make this syrup up to one day ahead. If you are doing it simultaneously with the pastry, make it first and put into the freezer to cool before using. Bring all ingredients to a boil, stirring sugar and water together until water is clear and sugar is dissolved. Be careful not to boil over. Maintaining a boil for about 5 minutes, test for syrup doneness by placing a small dot of syrup onto your thumbnail. If you can turn your thumb without the syrup displacing it's done—or if you have a candy thermometer, turn it off when it reaches the "syrup" stage.

Method for Pastry

Combine nuts, sugar, and cinnamon in a bowl and set aside. Carefully unroll thawed phyllo and place onto a piece of parchment paper near a 13x9-inch sheet pan. Make sure water from the sink or other liquids do not splash onto the phyllo, which will ruin the entire stack. Mix both olive oil varietals in a bowl, then oil the sheet pan—I use a 4-inch utility brush from the hardware store. Now use your thumb and separate the first sheet of phyllo from the batch, lift it up, and place it squarely onto the pan. Holding it down with one hand, make long, even strokes onto the phyllo with the pastry brush, covering 100 percent of each sheet as you layer one over the other, counting to 14.

Sprinkle half of the nut mixture onto the top layer. Oil 12 more sheets of phyllo to form a middle "filling" of phyllo pastry within the baklava, and sprinkle remaining nut mixture over this grouping. Top with final 14 sheets of oiled pastry. Cut the pastry into 1-inch strips down the length of the pan. Cut baklava into a diamond pattern by making 1-inch cuts on a 45-degree angle across the pan width.

Bake at 375˚, covered for 15 minutes, uncovered for 25–30 additional minutes, until evenly golden brown. Pour cool syrup over hot baklava. To serve, cut each piece again with a sharp knife to remove cleanly.

GREEK WALNUT CAKE WITH POMEGRANATE AGRODOLCE

This is a totally new take on my grandmother's recipe, which used butter instead of olive oil, very little baking powder, and a different combination and quantity of spices. Traditional *karidopita* (Greek Walnut Cake) is infused with hot simple syrup onto a cold cake. This cake works great with an Italian-style *agrodolce*—a traditional sweet-and-sour sauce in Italian cuisine. Its name comes from *agro* (sour) and *dolce* (sweet). Agrodolce is made by reducing sour and sweet elements, traditionally vinegar and sugar. I've chosen to make mine with Global Gardens Pomegranate Golden Balsamic Vinegar.

Ingredients

1 ½ c organic sugar

2 c sifted organic cake flour

3 tbsp baking powder

1 tsp allspice

2 tsp cinnamon

1 c ground organic walnuts

3 eggs

1 c milk + 1 tsp Global Gardens Champagne Vinegar

1 c Global Gardens Kalamata Extra Virgin Olive Oil

Method

Blend all dry ingredients with a fork. Mix in one egg at a time before adding milk, vinegar, then olive oil. A machine mixer works best to make sure all ingredients are blended together—don't overblend!

Bake at 350˚ in a ½ sheet cake pan or in an 11 x 15 ½-inch roasting pan for 35-40 minutes, until a toothpick inserted into the center comes out clean.

Cut cake when cold. While still inside the pan, pour pomegranate glaze over the cake. Place cut pieces into individual paper serving cups.

Ingredients for pomegranate agrodolce

2 c sugar

½ c Global Gardens Pomegranate Golden Balsamic Vinegar

½ c water

Method

Pour ingredients into a saucepan over medium-high heat, stirring sugar completely so that it dissolves into the liquid and does not burn. Bring to a rolling boil, being careful not to allow glaze to boil over the sides, until syrup makes a slight bead that will hold its round shape when dropped onto your thumbnail—the perfect "Greek" test for syrup! If you're more inclined, you may use a candy thermometer to the syrup stage; however, my yiayia's (grandmother in Greek) test makes the perfect consistency of syrup; sometimes candy thermometers make the syrup too thick for my liking.

CREDIT: ANITA WILLIAMS

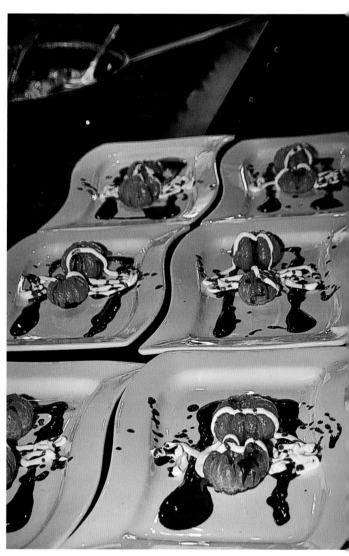

CREDIT: ROBERT DICKEY

CALIFORNIA TANGERINE DREAM

Crank up the old '60s music to the tune of "California Dreamin'," a tune I remember as a kid all too well. "On such a winter's day . . ."

Winter is the time for in-season tangerines to hit the stores (and your tummy), so here's a perfectly simple solution to the five-minute dessert challenge!

SERVES 4 (I LIKE TO USE CALPHA-LON OR A SIMILAR SKILLET FOR THIS RECIPE)

Ingredients

4 seedless California tangerines (satsumas are my favorite)

1 ¼ c organic 60% cacao chocolate chips

⅓ c + 3 tbsp Global Gardens Ascolano Extra Virgin Olive Oil

½ c crème fraîche sweetened with 1 tbsp organic California orange blossom honey

Method

Peel tangerines and stuff each core area gently (being careful not to break the integrity of the fruit) with approximately ¼ cup of the chocolate chips. Place into skillet oiled with 3 tbsp olive oil.

Cook

Roll tangerines carefully in 3 tbsp olive oil in a skillet over medium heat until white veins on fruit are clear, 3–4 minutes. Remove from heat.

In another nonstick pan, melt remaining chocolate chips over medium-low heat with ⅓ cup olive oil. Stir continually until chocolate is melted into a smooth sauce.

Presentation

Place tangerines on dessert plates with a creatively swirled drizzle of chocolate sauce. Garnish plate with 2 tbsp scoop of crème fraîche mixture.

PERFECT PECAN PIE

2 c organic sifted all-purpose or pastry flour

½ tsp salt

1 tsp vanilla or Grand Marnier (Armagnac or cognac work fine too)

⅓ c Global Gardens buttery Mission, Kalamata, or Koroneiki Extra Virgin Olive Oil, frozen into ice cube trays

¼ + 3 tbsp very cold half and half

Method

Mix the flour and salt together in a bowl with a fork. Place the bowl in the freezer for 20 minutes. Remove the bowl from the freezer and make a well in the middle of the dry mixture. Add the vanilla, olive oil, and half and half. Mix well with a fork, but do not overmix. Form the dough into a ball and refrigerate for 20–30 minutes. Then, roll the dough out onto wax paper, fold the dough lightly into quarters, and then center it in an ungreased pie dish. Carefully unfold the dough and create a decorative edge on the crust.

(Note: this recipe does not work well on a tin pie pan—glass or ceramic please.)

3 eggs

⅔ c organic sugar

½ tsp salt

⅓ c Global Gardens Mission/Manzanilla Blend Extra Virgin Olive Oil

1 c organic maple syrup

1½ c organic pecans, chopped

Method

Beat first five ingredients until blended, pour into crust, then add pecans; they will naturally float to the top and stay there when baked.

Bake at 350° for 40–50 minutes until middle of pie is set and does not jiggle.

LEMON-LIME & GINGER ICE CREAM

After a hot day in the Coachella Valley, the snowy San Jacinto Mountain Range called for us to explore. Hunting for wildflowers in the Sonoran Desert one minute, then taking the Palm Springs Aerial Tramway to Alpine Forest the next—it made for a perfect ice cream tasting opportunity and celebration ... all in one great Caliterranean day.

Ingredients

2½ c milk

1 pint of heavy cream

1 c sugar

¼ c each fresh lemon and lime juice

3 eggs, beaten

½ + ¼ c Global Gardens Ascolano Olive Oil

2 tbsp ginger juice

Cook

In a saucepan, whisk together and place on medium heat ½ cup of milk, ½ cup heavy cream, 1 cup sugar, ¼ cup lemon juice, 3 eggs, ¼ cup olive oil, and 2 tbsp of ginger juice. Continue to cook until the mixture turns a light yellow and has thickened.

Cool

Remove pan from heat, cover lightly, and place in refrigerator to cool long enough to add remaining ingredients without cooking, about 40 minutes.

Add

After cooling, add the rest of the cream, 2 cups of milk, ½ cup olive oil, and ¼ cup lime juice. Mix together well and place in an ice cream maker for 30–45 minutes until creamy.

Pairs perfectly with Solar Sauce (see page 202).

OLIVE OIL & VINEGAR FOR LIFE

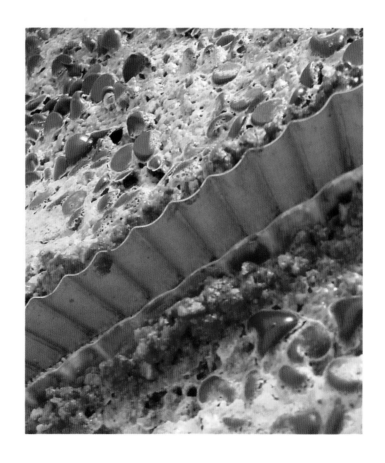

SMOKED SALT CHOCOLATE TART

Chocolate and salt have been a beloved combo of mine since I first moved out and used to think it was cool to eat a Hershey's bar with a bowl of potato chips for dinner. This dessert is a little more sophisticated and has a wider appeal. Santa Barbara pistachios are a great local food, although most of these nuts tend to come from the Middle East. Feel free to substitute almonds for the pistachios.

Ingredients for Crust

- 2½ c ground nuts (I use pistachios)
- ¼ tsp five-spice blend
- 1 egg, thoroughly beaten
- 2 tbsp Global Gardens Ascolano Extra Virgin Olive Oil

Method

Thoroughly blend all crust ingredients and chill for 30–45 minutes in the freezer. Press with a spoon into a 2-inch-deep round tart dish or deep pie pan. Refrigerate until ready to fill.

Make pecan pie filling on page 215, add 1 c 60% cacao chocolate chips to filling recipe and fill cooled tart crust.

Bake at 350˚ for 40–50 minutes until filling does not jiggle in the middle. Cool completely.

Ingredients for Chocolate Ganache Topping

- 2 c heavy cream
- 6 oz melted 60–72% dark chocolate
- 2 tsp Mendocino Sea Smoke Salt

Method

Whip together cream while simultaneously, slowly adding melted chocolate (so it will not curdle) with KitchenAid or other stand mixer. Spread over tart. Sprinkle with smoked sea salt.

MAR VISTA CRÈME BRÛLÉE

A high-energy day at Mar Vista Cottages preparing a gourmet meal for twelve people was brought to a tasty close with this signature crème brûlée. Cravings for this fine dessert can be accompanied by anxiety about its preparation. What if it doesn't set? What if it doesn't caramelize? Nonsense! This is easy to make and can handily be made a day in advance.

SERVES 6

Ingredients

1 qt organic heavy cream

1 tsp vanilla extract *or* 1 vanilla bean, opened and scraped (tastes richer!)

½ c white sugar

½ c brown sugar

6 egg yolks

1 tbsp globe basil, minced

1 tbsp grapefruit zest

about 2 qt boiling water

Method

Place cream and vanilla into a medium-size pan and bring to a boil over medium-high heat, stirring constantly. Remove from heat, cover, and set aside for 10–15 minutes.

Blend the two sugars together to form 1 cup sugar. Whisk ½ cup blended sugar and egg yolks rapidly until blended. Add cream, about ½ to 1 cup at a time, continuing to whisk thoroughly, adding basil and grapefruit zest last. Pour mixture into crème brûlée dishes and place dishes into a deep cookie sheet or cake pan. Pour hot water into the pan so that it comes halfway up onto crème brûlée dishes, being very careful not to get any water into the dishes themselves.

Bake at 325° for 40–45 minutes until center is firm but still just a little jiggly. Remove from pan of water and chill at least 4 hours before serving.

Tip: You can make this dessert up to 3 days ahead of time!

Reduce

Make about ½ cup reduction glaze from Global Gardens Pomegranate Balsamic Vinegar (see page 27 for instructions). Put reduction into Ziploc bag or small squeeze bottle. Make this at least 2 hours prior to serving and even up to a month ahead of time!

Presentation

Remove crème brûlée from the refrigerator about 20 minutes prior to torching, with remaining sugar sprinkled evenly amongst the 6 serving dishes. Melt the sugar using a butane torch until it is bubbly and crispy. Customize the top of the crème brûlée with a message, heart, or other festive symbol with reduction glaze.

CREDIT: RON BOLANDER

GLUTEN-FREE ALMOND SPA COOKIES

These popular cookies exceed every expectation in flavor and texture. You can keep them in the freezer for a couple of months, thawing when you want a shortbread-like utopian treat. I call them spa cookies because they are practically guilt-free (in moderation), and, quite honestly, I dreamed them up during an indulgent spa day while I was craving peanut butter cookies and shortbread. This recipe brings two of my all-time favorites together in a much healthier version.

Ingredients

2 c Bette's Gourmet Four Flour Blend or Premium Gold Gluten Free Flour (I've tried tons of flours for baking, and these two are the best.)

¾ tsp fine grain sea salt

1 c organic almond butter

½ c organic agave syrup

⅓ c + 1 tbsp Global Gardens Kalamata Extra Virgin Olive Oil

1½ tsp organic vanilla extract

1 c 60% cacao chocolate chips

3 tbsp ground organic raw almonds

Method

In a medium mixing bowl, combine the flour and salt. Add the almond butter, agave syrup, olive oil, and vanilla extract, then stir in the chocolate chips. Chill for about an hour. Drop by tablespoons onto ungreased cookie sheets, 2 inches apart, and sprinkle lightly with ground raw almonds.

Bake at 350 for 10–11 minutes. Let cool before removing from pan. Gluten-free has never tasted this spalicious.

ORANGE FANTASIA CAKE

A tribute to Walt Disney's brilliant production, the first commercial film ever to be released in stereo, I have created a symphony of flavors here that will surprise the most consummate dessert lover. Not supersweet, but with an excellent crumb and a gooey topping that tastes like . . . *what is that*? Surprises on your palate, in my opinion, are always a good thing— and just like Mickey Mouse juxtaposed to Stokowski, who doesn't like a surprise or two?

SERVES 8

Ingredients for Cake

- ¾ c cake flour
- 1 tsp baking powder
- ¼ tsp salt
- ½ c organic sugar
- 2 eggs, beaten
- 2 tsp fresh orange juice
- ½ c + 2 tbsp Global Gardens Blood Orange Extra Virgin Olive Oil

- 1½ tsp vanilla
- zest of one jumbo Valencia orange

Method

Combine dry ingredients and set aside. Combine eggs, orange juice, olive oil, and vanilla, mixing thoroughly at medium-high speed. Slowly add all dry ingredients, mixing until batter is well blended. Stir in zest.

Pour into 8 ½x4 ½-inch pan, very lightly oiled with a small amount of orange olive oil.

Bake at 350˚ for 45–50 minutes until cake is golden and inserted tester comes out clean. Allow cake to cool prior to glazing.

Note: For a fabulously light, gluten-free Orange Fantasia Cake, this recipe can be exactly duplicated by substituting Bette's Gourmet Four Flour Blend for the cake flour.

Ingredients for Glaze

- 1 c Global Gardens Blood Orange Dark Balsamic Vinegar
- 1 tbsp culinary lavender

Pour vinegar and lavender into a wrought iron or other heavy skillet. Bring mixture to a boil over medium-high heat and cook until liquid amount is reduced by 50 percent. Punch about 20 fork holes gently and evenly into cake. Pour hot glaze over cooled cake slowly, allowing it to absorb into the cake as you pour.

SMART TARTE TATIN

I'll admit it: this recipe came as a bit of a challenge. Rene Lynch from the *Los Angeles Times* shared her love for "real" tarte tatin as we were sharing a dish of California Tangerine Dream and a few other recipes, philosophizing about my recent olive harvest. Rene was pretty darned sure that tarte tatin *must* be made with butter. I was up for the challenge, and you can judge the results of this tarte tatin with olive oil for yourself. You can substitute pears for apples if you want to give your dessert a lighter twist. This was the first recipe where I tried freezing 1-tbsp measures of olive oil in ice cube trays—it worked perfectly!

Ingredients for the Crust

1 tbsp sugar

1¾ c all-purpose flour

4 oz Global Gardens 100% Mission Extra Virgin Olive Oil, frozen in 1 tbsp cubes

3 tbsp ice water

Method for the Crust

Using a food processor, combine the sugar and flour. Place bowl unit with blade, flour, and sugar intact into freezer. Remove from freezer after about 20 minutes and pulse olive oil cubes first to make a beady consistency, then add ice water, 1 tbsp at a time. Do not overprocess! Make a ball quickly and pat with flour if sticky. Use a floured rolling pin on a floured piece of parchment paper, rolling the dough into a 10-inch circle directly onto the parchment paper. Check crust to make sure it is big enough to cover the diameter of your skillet. Cover crust with another sheet of parchment paper and place gently into refrigerator.

Ingredients for Filling

6 medium apples (2 each Fuji, Gala, and Granny Smith)

Frozen cubes of olive oil!

3 tbsp buttery Mission, Kalamata, or Koroneiki Extra Virgin Olive Oil

1 c organic sugar

1 tsp fine sea salt

3 tsp cinnamon

½ c chopped raw almonds

Method for Filling

Peel and core the apples, cutting each one into fifths. Use a Le Creuset or similar ovenproof 10-inch skillet and stir olive oil, sugar, salt, and cinnamon over low-medium heat until well blended. Arrange apple pieces in a pretty spiral, starting with the outside edge, working your way into the middle. Sprinkle the almonds over the arranged apples. Any leftover apple slices will be used after cooking, prior to baking, since the apples will shrink in the first cooking step. Gradually increase heat and let mixture boil 5–6 minutes or until juices turn a golden brown color. Remove skillet from heat and carefully turn over the apples. Let cook again on medium heat for another 5 minutes.

Place crust over cooked apple mixture, tucking in the edges if it hangs over the side of the skillet. Make about 12 fork holes around the entire crust, not just on the outside edges.
Bake in the skillet at 375˚ for 25–30 minutes until crust is golden. Remove from oven and cool for 30 minutes, but not completely! Glide a sharp knife along the edge of the skillet to loosen the crust and any carmelization from the sides of the skillet. Place serving plate over top of skillet and quickly flip over so that tarte tatin drops down onto the dish. Be careful because the skillet will be hot, heavy, and unwieldy, but you can do it! Leave the skillet upside down on the serving plate for about 10 minutes to ensure that everything has fallen down onto the crust. When you remove the skillet, place any filling or apples that have stuck to the skillet directly onto your finished tarte tatin. Now, isn't that smart? One taste and you will know that to be true.

Caliterranean

PET TREATS

Nearby fields planted with alfalfa, corn, sweet canning tomatoes, pumpkins, and zucchini make fun romping grounds for GiGi and Zac. Creating just the right treats for them was tough, as they had so many wild options out their back door. But photographing them was more fun than coming up with recipes for treats. GiGi was such a good girl—I had her pose with a squirrel-shaped treat up against her paw. She gave me a quick glance as if saying, "She really expects me *not* to eat this incredible home-baked treat?" She gobbled it up happily when I gave her the okay. Zac, on the other hand, had to burn off some energy playing in an empty fifteen-gallon plant container before he calmed down enough to cooperate properly. Anita, our photographer, decided to just go ahead and shoot his photo inside the black tub—a handsome contrast!

HAPPY DOG MUNCHIES

Half Anatolian, half Great Pyrenees, GiGi is our mascot at Global Gardens, and justifiably so! These treats have been great inducements to learning new tricks.

Tip: Freeze all but a few days' worth after making this recipe. Because they are made with fresh products, they will spoil after a few days in the fridge. Even frozen, they break apart easily and are gobbled up longingly.

Ingredients

1 lb ground lean meat (any)

2 ½ c organic whole wheat flour

½ c powdered egg whites

1 tsp salt

1 egg

½ c + 2 tbsp Global Gardens Greek Koroneiki Extra Virgin Olive Oil

¼ c water

Method

Cook meat thoroughly in a skillet. There should be absolutely no fat in the skillet when combining with all the ingredients above into a bowl. Mix well and roll out into 3/16-inch-thick circles. Use cookie cutters to cut fun shapes.

Bake on ungreased cookie sheet at 350° for about 20 minutes until golden brown on the bottom.

CALI KITTY CRUNCHIES

We live in the country, so it's Zac's honor-bound duty to monitor the property for any errant rodents who might be running along the creek or in the field of tall grasses next door. Having to wait for a treat was something he had never experienced—he's always bringing us *his* treats!

YIELD: 150 TREATS

Ingredients

- 6-oz can drained tuna, packed in springwater
- 1 c cornmeal
- ½ c rice flour
- ½ c whole wheat flour
- ½ c Global Gardens Greek Koroneiki Extra Virgin Olive Oil

Mix all ingredients together thoroughly and roll out to approximately ¼-inch thickness to fit onto a cookie sheet. Using a pizza cutter directly on the pan, cut dough into approximately ¼-inch cubes. Nobody is going to measure them, and your cat won't know the difference if you cut crooked, too thin, or too fat. Watch as love at first bite ensues!

IN CLOSING

The possibilities for living a Caliterranean life are endless. This cookbook focuses on using Global Gardens Extra Virgin Olive Oils and Balsamic Fruit Vinegars in every meal of every day . . . for life! Well-being of body and mind is a significant component of the Caliterranean lifestyle one can choose to have anywhere, anytime, forever. I hope you will join me and my family on our Caliterranean journey for many years to come.

IN GRATITUDE

To my daughters Anita and Sunita, who learned how to wait until photography sessions were over to eat dinner together. (Anita, thank you for shooting some great pix for this book, and Sunita, for suggesting new flavors and food styling expertise; you are both treasures.) Special thank you's to my sister Renee Nicholas and to Sidney Schultz—the first to believe in Global Gardens—now to Bill Palladini, who continues to do so. My sister (poulaki) Kathy is always there for me. Nancy Garvin is a dear friend who believes without falter *and* makes a great camping/cooking buddy. Daniel Lentz generously created a few recipes in this book that I varied only slightly; his food talk, writing, and music are munificent art. My entire staff at Global Gardens is, without question, beyond perfection; without them this project could never have been completed with such joy and immersion. Jenn McCartney, my editor at Skyhorse, was the first to hear—and like—the word Caliterranean; her patience with me is humbling. Thank you, Tony Lyons, for saying yes, and Tad Crawford, for your steadfastness throughout the years.

CREDIT: DORI BASS

I invite you to come visit my latest creation, the Global Gardens Los Olivos, California Farm Stand, founded in 2014, where I love to provide what I honor as "taste education." Here I have created a comprehensive Tasting Bar, where I guide you through the different flavor profiles of my Extra Virgin Olive Oils and Fruit Balsamics profiled in this book. We'll even taste olives and olive oil directly from this 3-acre retreat in Santa Barbara wine country. Oh . . . I also love to teach cooking classes and healthy living the Caliterranean way—at your place or mine—just email me for more information!

Enjoy shopping at my online store for extra virgin olive oil and fruit balsamics (plus about twenty other food products I make!): globalgardensonline.com.

Please e-mail me with questions, thoughts, and recipes:

theo@globalgardensonline
Follow Theo@caliterranean on Twitter.